Algebraic Models for
Accounting Systems

Algebraic Models for Accounting Systems

Salvador Cruz Rambaud
José García Pérez
University of Almeria, Spain

Robert A Nehmer
Oakland University, USA

Derek J S Robinson
University of Illinois at Urbana-Champaign, USA

World Scientific

NEW JERSEY · LONDON · SINGAPORE · BEIJING · SHANGHAI · HONG KONG · TAIPEI · CHENNAI

Published by

World Scientific Publishing Co. Pte. Ltd.

5 Toh Tuck Link, Singapore 596224

USA office: 27 Warren Street, Suite 401-402, Hackensack, NJ 07601

UK office: 57 Shelton Street, Covent Garden, London WC2H 9HE

British Library Cataloguing-in-Publication Data
A catalogue record for this book is available from the British Library.

ISBN-13 978-981-4287-11-1
ISBN-10 981-4287-11-3

Printed in Singapore.

"The imagination ... gives birth to a system of symbols, har-
monious in themselves, and consubstantial with the truths of which
they are the conductors."

Samuel Taylor Coleridge, "The Statesman's Manual", 1816.

Preface

In recent years there has been no shortage of applications of mathematics to economics, mainly through the use of methods from statistics, probability and risk analysis. It is much harder to find significant applications of abstract algebra to the area. However, the rise of the information sciences has clearly displayed the opportunities for applying what used to be considered the purest of pure mathematics. It is now commonplace for students of computer science to take the time to acquire a basic knowledge of algebra. It is not hard to see why algebra should be enjoying popularity: in the analysis of complex systems of all kinds the power and precision of algebraic concepts, and sometimes just algebraic notation, can be an enormous aid. On the other hand, one can search the literature in accounting theory and find few attempts to make use of algebra, and what there is tends to be at quite a modest level.

The object of the present work is to make the case for applying algebra to the study of accounting systems by finding algebraic concepts which are able to reflect accurately the workings of real life systems. The benefits of such a study are diverse: the demand of algebra for precision compels us to question and make exact everyday ideas and processes in order to express them in abstract form. It also serves to provide tools to analyze accounting systems.

The concepts which appear most frequently in the present study are: column vectors with zero sum, the so called *balance vectors*, which reflect the perfect balance of an accounting system; *directed graphs* to show the flow of value through the system; *automata* to model the computational aspects of accounting. These in turn lead to further algebraic concepts such as monoids, subaccounting systems and quotient systems. All of these notions provide valuable ways of looking at and understanding the operation of accounting systems.

In a further departure from previous attempts to inject algebraic ideas into accounting, we emphasize the role of rigorous proofs – this is, after all, the only way to achieve any kind of certainty. In addition, where it seems of mathematical interest, we have not hesitated to follow up on mathematical questions that are suggested by accounting concepts, sometimes in the form of combinatorial problems.

The current work had its origin in the Ph.D. dissertation of the third author at the University of Illinois at Urbana-Champaign in 1988 and in a subsequent article in collaboration with the fourth author. In addition, the introduction of automata into the study of accounting owes much to an article by the first two authors. This book is a greatly expanded version of these works. The first chapter contains an extended account of previous approaches to accounting theory by diverse authors, the object being to provide a setting and historical background for the current enterprise.

While every effort has been made to keep the book self contained, inevitably it is necessary to assume that the reader has a certain level of mathematical sophistication, roughly what one would expect of a student who has taken at least a first course on discrete mathematics. However, abstract structures such as monoid and automaton are fully explained: a reader who would like to have more background in abstract algebra should consult one of the innumerable texts on the subject, for example [7] or [8].

The authors are grateful to the Department of Mathematics at the University of Illinois, and Ms. Sara Nelson in particular, for assistance with the technical typing. The last author thanks the University of Almería, Spain for exceptionally warm hospitality during a visit there in June 2009. Finally, the authors thank Ms. Tan Rok Ting of World Scientific for her able assistance at all stages of this project.

Contents

Chapter One

Approaches to Accounting Theory

"Perhaps I am busied with pure numbers and the laws they symbolize: nothing of this sort is present in the world about me, this world of 'real fact.' And yet the world of numbers is also there for me, as the field of objects with which I am arithmetically busied; while I am thus occupied some numbers or constructions of a numerical kind will be at the focus of vision, girt by an arithmetical horizon partly defined, partly not; but obviously this being-there-for-me, like the being there at all, is something very different from this. *The arithmetical world is there for me only when and for so long as I occupy the arithmetical standpoint.*"

Edmund Husserl, Ideas p. 94 (italics original)

1.1. Historical Perspectives

Accounting is an ancient human activity. From the time when men and women first engaged in trade, whether for barter or money, it must have been necessary to keep some kind of record of incomings and out-goings, to which the origins of the double entry bookkeeping system can be traced. Already in the twelfth century of the Christian Era the Arabic writer Ibn Taymiyyah mentioned in his book Hisba (literally, "verification" or "calculation") accounting systems used by Muslims as early as the seventh century. A critical development in the history of accounting was the publication in Venice in 1494 of the book "Summa de Arithmetica, Geometria,

Proportioni et Proportionalita" by the Franciscan monk and mathematician Luca Pacioli (1445-1517) – see Pacioli [1963]. This is the first known work to contain a detailed description of the practice of bookkeeping and the double entry system, "Particularis de Computis et Scripturis". Today it is widely regarded as the forerunner of modern bookkeeping practice. It was also Pacioli who introduced the symbols for plus and minus, which became standard notation in mathematics during the Renaissance. The first book on accounting in the English language appeared in London in 1543, authored by John Gouge. An important source for the early history of accounting is the writings of R. Mattessich ([1998], [2000], [2003], [2005b]).

While it seems clear that accounting was considered by Pacioli and his contemporaries to be part of arithmetic, its relationship with other parts of mathematics has had to wait much longer for recognition. The methods of statistics have long been used, almost since the importance of that branch of applied mathematics was first recognized in the seventeenth century. More recently probability theory and risk analysis have featured in economics. However, algebra has played little or no role, despite the precision of its language and its ability to describe complex situations concisely. The purpose of this monograph is to draw attention to the contribution that abstract algebra can make to accounting theory. Indeed it is the authors' contention that, at least in its deterministic form, accounting theory should be considered as a branch of applied algebra.

The book presents and develops a proof-based, algebraic approach to the study of accounting systems. The analysis provides a description of single firms in terms of abstract algebraic objects such as automata. It concentrates on the process of producing information from data provided by the environment through the double-entry system. This process, although considered by many to be the core of accounting, has often been ignored in accounting research. In attempting to address this issue, the book adds a level to the analysis of the information economists through the very act of exploring the production aspect of accounting information systems. The motivation is to expose the complexities and subtleties of information production in this field of research. The literature review which follows reflects the rather fragmented nature of the work which has been done up to this time in axiomatics, natural languages, formal grammars and information economics. The book shows how a basic accounting system can be represented as a formal algebraic language. The reduction of accounting systems to these types of languages will lead to a much stronger method of modeling information systems.

Although much discussion has occurred in the last fifty years concerning the treatment of accounting as a language and its justification as the language of business, surprisingly little progress has been made. This is perhaps due to the remarkable diversity of methods in linguistic research. In pure linguistic research, the various methods are divided into the natural language and formal language schools. The natural language schools study naturally occurring human languages as they have arisen from the historic acts of increasingly complex human communication. The formal language school arose from this tradition as methodologies were devised to study natural languages. These methodologies generally tried to reduce the complexity of natural language constructs to a finite system of grammatical rules. The formal language school became distinct from the natural language school when it was determined that certain domains of language, such as parts of mathematics and later computer science, could be completely specified by these finite systems.

Outside the area of pure linguistics, some applied fields such as speech communication and organizational behavior have adopted certain linguistic approaches and have developed other approaches independently. Semiotics has been used to determine what signs employees attend to in their everyday work relationships (Barley [1983]). Semiotics studies the meanings that people assign to language constructs in their search for understanding in their worlds. Recently, hermeneutics has been used to develop a criticism of the economics literature (McCloskey [1983]). This method employs the analysis of texts to identify repetition of linguistic constructs or changes in constructs over time and to study how the authors of the texts view their social realities.

With such a diversity of methods available, it is hardly surprising that the accounting profession has found little success in its search for a formalization behind the intrinsic meaning of the metaphor "accounting as the language of business". It is the contention of the present writers that the best way to proceed in the issue is to choose a potential methodological candidate, develop it and make a judgement based on its contribution to accounting research. The method chosen here is a formal, algebraic approach. In order to present this new approach to accounting in its contemporary setting, a detailed review of the language studies, both formal and natural, which have appeared in the accounting literature up to this point, is given in the sections which follow.

1.2. Algebraic and Proof-Based Approaches

As has been pointed out, the application of abstract algebra to accounting is something of a novelty. However, it would be wrong to suggest that nothing has been attempted in this direction. Already in 1894 the English algebraist Arthur Cayley wrote that "The principles of book-keeping by double entry constitute a theory which is mathematically by no means uninteresting; it is in fact, like Euclid's theory of ratios, an absolutely perfect one, and it is only its extreme simplicity which prevents it from being as interesting as it would otherwise be" (Cayley [1894]). Even before this time matrices had been introduced in the framework of accounting theory by Augustus De Morgan [1846], a route that was not followed by other writers until 100 years later. Indeed matrices reappeared as a topic of research interest in accounting only in the 1960s and 1970s, when a number of classic works in accounting theory were published, such as Edwards and Bell [1961], Chambers [1966], Ijiri [1967] and Mattessich [1964]. Here it should be understood that matrices were considered only as a tool to describe in a mathematical way the activity of accounting, and not as an attempt to formalize the concept of an accounting system. Paton [1922], one of the major personalities in accounting research in the United States in the 1920's, seems to have been the first author in formulate some accounting postulates. Nevertheless at that time fundamental research was not common in this area and the postulates never became part of a formal system.

Perhaps the most famous axiomatization of accounting was given by Mattesich [1957, 1964]. The first of these publications relies on a matrix formulation of accounting to provide structure to the axiomatic system. Three axioms are included in this schema: a plurality axiom, a double effect axiom and a period axiom. The first asserts that there exist at least two objects with a common measurable property. This provides a basis for the recording of transactions. The second axiom states the existence of an event which causes an increase of a property of one object and the corresponding decrease of the property of another. In effect this is an axiom of double entry. The last axiom requires that accounting systems are capable of being divided into time periods, thus providing a basis for the construction of financial statements. In addition to these axioms, the paper provides numerous definitions and "requirements", as well as several theorems.

The proofs of the theorems in Mattessich's first paper give insight into the formal relationship between the axioms and the theorems. The proofs consist of algebraic manipulations of matrices using the sigma, i.e., summation, notation. While in a sense this does serve to "demonstrate" the theorems from the definitions, the proofs do not consist of formal deductions from the axioms, as would be the case in a strict deductive system. Thus the axioms do not serve as a complete basis for the proofs of the theorems. In his second publication Mattessich shifts from a matrix to a set theoretical approach. In this work he relies on primitive terms, definitions stated using the set notation, and propositions. The theorems which are proved appeal to the definitions and propositions and are basically algebraic in nature. Perhaps the absence of axioms in this second work was due in part to the difficulty noted above, i.e., axioms which are not used in the proofs of the theorems. Some might argue that the propositions substitute for axioms in this formulation, but the propositions here are generally set theoretic definitions of such concepts as an accounting period or the chart of accounts. Although they may be invoked as a proof proceeds, the proofs do not begin with the propositions, nor are the theorems deduced from them. Again the beginnings of a formal proof-based system can be discerned here, but it is not coupled with a formal deductive scheme. This type of scheme may be provided by including the axioms of the mathematical system – in Mattessich's case an algebra – as part of the axiom scheme, thereby specifically allowing for mathematical inference within the axiomatized system, as will be seen below.

Ijiri's [1975] book on accounting measurement also includes three axioms, but again it lacks any derivation of theorems from the framework of these axioms. He does, however, derive his axioms from the theoretical structure of the accounting system which he provides in the book. Therefore it is likely that he sees these axioms more as general statements about accounting, rather than as a basis for any formal deductive system. Indeed he makes no attempt at all at proving the theorems. One of the contributions of the current book is that it provides not only an axiomatization of accounting systems, but also a deductive inference scheme which can operate on the axioms in a formal way to derive the theorems as consequences.

Tippett [1978] derived axioms of accounting measurement, and more recently Cooke and Tippett [2000] used a structural matrix to represent the restrictions imposed in a double-entry bookkeeping system, employing the information in the matrix to predict financial ratios. Willett [1987, 1988] demonstrated in two papers the derivation of axioms of accounting measurement, following Tippett's methods. His analysis extended to the stochastic space of accounting variables. Gibbons and Willett [1997], building on Willett's earlier work, demonstrated that accounting data produced from implemented information systems have a statistical nature due to the error generated by processing: that statistical nature is shown to be of value to decision makers under certain conditions. Nehmer and Robinson [1997] provided an initial description of the algebraic structure of accounting which is greatly expanded upon in this book. Nehmer [2010] encodes the algebraic structure in first order logic and derives consequences for the resulting structures.

Aukrust [1955, 1966] made an important contribution to the standard methodology for national and international accounts, completing a theoretical discussion of the underlying principles in accounting at the national level. He presented some problems of definition, classification and measurement of national accounts in an axiomatic way. After stating a set of twenty postulates, he showed that the structure of a simple system of national accounting can be derived from them. In this way it is possible to establish algebraic relations among national accounting concepts. Aukrust concludes: "The set of twenty postulates used above to derive a national accounting system is, of course, not the only one which could be conceived of. Others are equally feasible. Some would lead to national accounting systems different from the one described here, in much the same sense as non-Euclidean geometries are different from Euclidean geometry".

The problem of financial statements was dealt with by Arya et al. [2000], emphasizing the power of the double entry system to determine all consistent transaction vectors. They showed how a graphical representation of the accounting system can be used to obtain the characteristics of the vectors, solving in a simple way the problems of inverting and selecting the most likely transaction vector from the set of consistent transaction vectors. Arya et al. [2004] provided a systematic approach to reconciling diverse financial data. Again the key is the ability to represent the double entry system by a network of flows. Two specific uses are investigated: the reconciliation of audit evidence with management by means of prepared financial statements and the creation of transaction level financial ratios.

The first collaboration in the area between a philosopher of science and a theoretical accountant materialized in Balzer and Mattessich [1991, 2000]. They considered the reconstruction of yield to be a viable way of capturing the essence and basic structure of accounting as rigorously as possible. The proposed reconstruction showed that accounting has the same overall structure as other empirical theories by presenting nine axiomatic principles to establish the following concepts: economic objects, economic transactions, state-space for accounting, accounting data systems, accounts, double entry accounting systems, accounting morphisms and accounting systems (in general). By combining these definitions, they obtain the kernel of a model for accounting and they claim that all special methods and procedures used by accountants can be obtained from this core model with some appropriate specifications. All theorems are proved, but the authors indicate the need for further development of the axiomatic system presented in the paper and they present details of certain specifications to appear in future work.

According to Ellerman [1982, 1985, 1986], "Double-entry bookkeeping illustrates one of the most astonishing examples of intellectual insulation between disciplines, in this case, between accounting and mathematics". He described a mathematical basis for a treatment of double-entry bookkeeping in terms of the so-called "group of differences", sometimes called the *Pacioli group*: for details of this connection see 3.1 below. The possible use of the algebraic concept of a group in accounting theory is also considered in Brewer [1987] and Botafogo [2009], but with little progress beyond the formulation of some definitions.

There have been many other attempts to formalize accounting in a scientific way. Since the present work does not pretend to give an exhaustive history of accounting, only some of them have been mentioned. Details of other attempts can be found in Mattessich [1995, 1998, 2000, 2003, 2005a, 2005b].

On a final note, recently Demski [2007] has tried to answer to the question "Is accounting an academic discipline?" After analyzing the meaning of "discipline" and "academic", his immediate conclusion was negative. However, Demski was not pleased with this answer and therefore he preferred to analyze the ten indicators of the accounting as an academic discipline, ending with "... accounting is not today an academic discipline; it is an ever-narrowing insular vocational enterprise. But it could and should, in my opinion, be an academic discipline. Even if you disagree with my assessment, you should consider whether the state of academic accounting is, in your view, what it could and should be. The stakes in this game are enormous and serious".

1.3. Natural Language Approaches

Research in accounting as a natural language, as opposed to an proof-based system, has fallen into three broad categories: connotative and denotative meanings, readability of reports and linguistic relativity (McClure [1983]). The connotative and denotative meanings of language refer to its subjective and objective meanings respectively. The research in this category has emphasized the interpretation of accounting concepts by different groups including certified public accountants (CPA's), users, students and academics. The results have generally indicated agreement on the connotative meaning between groups, but there is some evidence of disagreement over denotative meanings (Belkaoui [1980b]). Research into the readability of financial reports has stressed the ability of the reports to communicate information on several levels. Levels of reading ability needed to comprehend the reports have been tested, but the tests were found to be inappropriate for the analysis of materials in a report format. Lebar [1982] tested several different types of financial report along an extentional - intentional axis. Extentional language is more descriptive and objective, whereas intentional language is more general and unqualified. She found that 10-K reports (a specific type of filing that a company makes to the Security and Exchange Commission) scored well on the extentional components as compared to the annual reports.

The third category of linguistic research in accounting is based on linguistic relativity (the Sapir-Whorf hypothesis). The two basic concepts of the hypothesis are that language determines thought and that consequently individuals with different linguistic backgrounds have different world views. Belkaoui [1978, 1980a] used this hypothesis to study disclosure issues in the area of pollution control costs, with results generally supporting the hypothesis. All three categories of research in accounting as a language have viewed it as a natural language and applied natural language techniques to its study.

Some more recent studies of business communication include Tyrvainen et al. [2005], who examine the internal and external communication of three business units, looking at digital, paper-based and oral communication. In a series of articles in the accounting area, Fisher [2004], Fisher and Garnsey [2006] and Garnsey and Fisher [2008] codify the professional accounting literature. This codification is then used to critique the adequacy of the literature (Fisher [2004]) and to examine amendments to the literature (Fisher and Garnsey [2006]). Garnsey and Fisher [2008] implement a software retrieval solution to the professional accounting literature.

An alternative approach is to view accounting as a formal language built up from a detailed specification of its grammar by exact rules of composition known as production rules. Formal grammars and languages were originally developed for natural language research and are still used there, especially in computational linguistics research. They have been largely absorbed into computer science because they are an alternative representation of finite state automata. Such automata are used in computer science for the general representation of computer languages. The concept is easy to relate to for anyone who have ever tried to learn a computer language with its peculiar sentence structure and rules. A good example of research using the automata approach is Cruz Rambaud and García Pérez [2005].

Demski et al. [2006], looking for a new language for the treatment of accounting information, examined the nature of quantum information in order to search for promising conceptual applications to accounting. They present some important features of quantum information such as quantum superposition, randomness, entanglement and unbreakable cryptography, and they begin to explore the possible link between quantum information and double-entry information which lies in the core of accounting information. The starting point is the work of Cayley [1894] on the parallel between the Euclid's theory of ratio and the double entry theory. As a consequence, it is intended to explore the possibility of a hybrid between accounting information and quantum information, "quantum double-entry information". In a second article, Demski et al. [2009] studied the applications of conceptual topology to quantum information and accounting information. The use of topology allows one to emphasize the qualitative characteristics of accounting information and to maintain the quantitative ones.

A reasonable and effective mathematization and axiomatization of the economy, and in particular of accounting, necessarily implies Diophantine formalisms (Velupillai [2005]), which raises issues of undecidability and non-computability. In the future there should be greater freedom for experimental research supported by alternative mathematical structures. In conclusion Velupillai speaks of "the notion of a Universal Accounting System, implied by and implying Universal Turing Machines and universality in cellular automata".

1.4. A Formal Grammar Approach

One exception to the exclusive use of natural language research methods in accounting is Stephens, Dillard, and Dennis [1985], hereafter Stephens et al. The article is entitled "Implications of Formal Grammars for Accounting Policy Development" and it presents a classification scheme for proposed and existing Financial Accounting Standards Board (FASB) statements. The examples provided in the article are partial formal grammars, reflecting the accounting rules promulgated by a specific standard. The level of analysis is macro in the sense that it considers the standard for all firms to which they apply. As such the analysis focuses on establishing criteria with which to evaluate standards through formal grammars. The three criteria used are possibility, consistency and resolution.

Possibility refers to the ability to reduce the statement to a formal grammar. One potential problem here, albeit one which is discussed in a different section of the article, is the difficulty in determining the primitives of the grammar. In the article the example of leases is cited. The determination of whether a certain economic event should be classified as a rental arrangement or a purchase has become increasingly problematic in accounting. Unless a clear demarcation is allowed or imposed on the "correct" interpretation of such an event under every circumstance, the formal grammar will not be capable of operating in these types of situation.

The second criterion, consistency, refers to the cross-statement compatibility of the grammars. This compatibility can perhaps best be addressed in terms of first order logic, rather than the formal grammar approach used in the article. The two systems are equivalent, so the change in approach is warranted. In first order logic consistency is defined in terms of the sentences which can be proved from the axioms. If both a sentence and its negation are provable from the axioms, then the system is inconsistent and in fact any sentence is then provable from the axiom system. In terms of the article, in order for a formal grammatical analysis to succeed, a single formal grammar containing all accounting standards must be demonstrated. Then any proposed new standards could be appraised in terms of their consistency with the current formal grammar.

The third criterion, resolution, is an attempt to deal with problems of inconsistency arising from the different rules specified in the single formal grammar mentioned above. The article proposes that uniform ranking rules be included in the grammar in order to remove such inconsistencies. It points out that the FASB does provide such rules in certain situations, but that the rankings so provided have not been uniform in the past. Stephens et al. classified the resulting inconsistencies as being due to one of three situations: arbitrary selection among possible standards, stipulation of standards without theory and the inability to write a definitive grammar.

In the first situation a choice is made and a particular standard must be selected, when alternative standards have possible correct economic interpretations and their own supporters. Stephens et al. contend that this and the next situation result primarily from lobbying by factions of the accounting community. The next situation occurs when a standard is stipulated which is lacking in theoretical support; this seems to mean lacking in terms of a justifiable economic interpretation. The interpretation is usually only provided a posteriori and may be thought of as imposing a new economic reality based on the standard. The last situation is the problem of specifying the primitives of the grammar, which was discussed under "possibility" above.

Stephens et al. divide the economic realm into three parts, the environment, accounting and decision. The effects of economic events in the environment are actions which play the role of primitives subject to the grammatical rules of accounting. The rules produce accounting results which are used by decision makers to produce decisions. Stephens et al. restrict their analysis to the accounting component only, so that the evidence of a transaction occurring is taken as a given and the use of the output is not analyzed. The same position is adopted throughout this book.

However, there are several differences between the article by Stephens et al. and this book, perhaps the most important being the level of analysis. The analysis presented here is at a micro level, as opposed to the macro level of the article. Specifically the analysis here pertains to the accounting system of a single firm. Secondly, a complete axiom system is developed for the firm, based on the double-entry components of the system only. The necessity of developing such a restricted system is based on the requirement of demonstrating the existence of such representations of accounting systems before proceeding with higher level analysis, as is recognized by the authors of the article.

A contribution of this research is to provide a basic method for constructing formal proofs in accounting. It interfaces with the axiomatization and formal inference scheme to yield a formal abstract specification; this leads directly to axioms for an accounting system, as well as to a system of inference which can be used to derive consequences of those axioms. In fact, the analysis of the paper includes the consideration of information systems as finite state grammars (FSG's) and automata. This representation is the basis of the computer languages which form the structure of any computerized system. Therefore finite state grammars can be used as a general representation of the process involved in converting states into signals. Such FSG's include relation as well as function operators, thereby providing a more powerful means of analysis in exploring the possibilities and limitations of the signal/output generation process of information systems.

The representation of information systems as FSG's serves two purposes in this analysis. First it addresses some problems noted below with information economics methodology, i.e., it provides a specific formulation of the internal production of information and assigns a specific interpretation to the states recognized by the system as well as its outputs. Furthermore, it allows for the production of multiple derivations from the capture of an additional piece of data. The second use of FSG's is to provide a convenient bridge between the representation of accounting systems as FSG's and their representation as proof-based systems. This is accomplished through the conversion of the production rules of the FSG into axioms of a first order logical system.

1.5. Information Systems in Information Economics

This book addresses\some of the issues in the comparison of information systems which occur in the information economics literature, this being the current standard of comparison of systems in accountancy. A large body of work has been done in the area using utility analysis and relying on the results of Blackwell's "Comparison of Experiments" (Blackwell [1951]). As the title indicates, Blackwell's procedure shows that if an experiment A is a sufficient procedure for a different experiment B, then A is more informative than B, i.e., it provides at least as many statistical measures. Authors such as Gjesdal [1981] have used the matrix form of Blackwell's results to analyze different information systems. Demski [1980] and Demski, Patell and Wolfson [1984] have used the basic matrix framework of states crossed with signals and in the latter paper relied on Gjesdal's information systems comparison result. All of these comparisons of information systems are founded on the partitioning of the states of nature, the idea being that different information systems will be able to "recognize" different states at various levels of fineness. That is, a certain information system may produce signal Y_1 when it recognizes S_1 and signal Y_2 when it recognizes S_2, whereas another information system may not be able to distinguish S_1 from S_2, and produce the same signal for either realization.

The implication of the states to signals model of information systems is that there is a set of functions corresponding to the set of information systems under comparison. Mathematically the conversion of the states to signals is a mapping from the set of possible states to the set of possible signals. Over the entire state and signal spaces the function family is neither injective nor surjective. In the first place a particular information system function may map two or more states to the same signal, so the function is not injective. Secondly, an information system function may not be able to generate certain signals in the codomain at all, so it is not surjective. Indeed in Demski's 1980 examples, it is only in the perfect information case that the mapping can be bijective, i.e., both injective and surjective. It is this lack of uniformity in the construction of the state to signal functions (or information systems) regarding their relevant domains and codomains which partially explains the failure of Blackwell's comparison technique in proof-based systems.

Several topics are important to the present analysis. Firstly, Blackwell's result lies in the domain of experimental procedures, whereas an information system is, in a practical sense, an extant structure generating outputs from inputs by a formalized system of rules. As such there are several differences in the level of analysis which are apparent. Most importantly the information economics analysis considers the external or environmental states of nature only, without considering the internal states of the information system itself. It therefore ignores the interaction of the internal components of the system in the production of its outputs.

The unspecified nature of the internal components prevents the methodology from addressing questions relating to how changes in the configuration of the system will alter the signal set generated. Of course, information economists use the term "information system" in a different sense than is used here. But it is the difference in representation of the system which allows this additional analysis to occur. These are important questions for the accounting profession since they involve the production of information for decision makers in an organization from the design of the accounting system.

This lack of concern for the internal state of the system also forces the information economics methodology to ignore explicitly the problem of data capture versus information production. That is, a state may occur in the external environment which is captured or recognized by the system but is not processed in a timely manner. While the techniques of information economics do implicitly take this into consideration by collecting states into sets based on the concept of fineness, this does not help in determining why a particular output is not being generated, i.e., whether the data are being processed too slowly or are not available at all.

A second deficiency in the statistical analysis of information systems is its inability to recognize that a particular state may generate more than one signal. The information economics approach provides, at best under perfect information, a single state or a single signal mapping for output production from information systems. Under imperfect information several different states may produce the same signal, but the reverse situation, of a single state being mapped onto multiple signals, is not considered. For instance, a decline in interest rates may cause changes in pension funding requirements, a decline in the mortgage interest rates being paid by an organization and declines in the dividend rate expected from an investment in mutual funds. Further, it is possible within an information systems methodology to develop single states to multiple signals if a recognition of the interrelationships between states and signals is provided.

A final problem with the current method of analysis is that it does not provide a convenient way to interpret the states and signals. As an example consider Gjesdal's [1981] description of an information system. Here he reduces the system to merely the specification of the signal's functional form and proceeds to assert that "the nature of the signals is of no concern" (p. 212). One can only assume that the nature of the information system is of no concern as well, yet it is difficult to comprehend the purpose of comparing objects whose nature is not the object of comparison.

Generally the matrix representation of information systems and especially the concept of state (and hence signal) partitioning does not address the problem of how the signals are generated. This leaves open the question as to whether and to what extent in the context of axiomatic information systems, such a partitioning is possible. This problem is addressed in Chapters 2 and 3 where the algebraic core of the model is constructed.

Demski's ([1980]) analysis of information systems differs from an axiomatic approach in that his complete model consists of a set of acts, states, state probability functions and utility functions, with states and acts as parameters. This model is conditioned on the decision makers' experience and assumes that the four factors mentioned above are correctly specified. He presents two cases, the perfect and the imperfect information situations. Under perfect information, the decision maker can directly observe the realization of the states of nature. Therefore there is no need for the information system to produce signals relating to the acts of an agent. Since the state is known with certainty before an act is chosen, this situation will match well the derivations of an axiomatized information system. If the state is known for certain prior to the act, there must be some decision procedure which would indicate which state will occur and such a procedure is axiomatizable. This is the case because under state certainty the state must already be a fact, in which case its truth value is known or must be determinable under some formulation which perfectly correlates its own predictions with the actualization of those predictions. In the latter case, the decision procedure will be reducible to a first order formula, barring the serious consideration of some form of crystal gazing as providing perfect information. Of course Demski would agree that this is an unlikely situation in any complex decision making problem. In fact, testing which state will occur is likely to involve a great deal of computational complexity in a complex, decision making environment: the results, if they can be determined with certainty, may not be produced in a timely manner.

In the imperfect information situation the information system cannot necessarily distinguish each state uniquely, so the same signal or output from the system may occur after the realization of different states of nature. When the state is not known with certainty prior to the act, Demski posits the information system as producing a set of signals which may, but usually do not, indicate the state which was achieved or which has transpired. The signal is a function of the information system with the states as the input and the signals as the output.

One and only one signal is associated with each state occurrence, although the same signal may be produced by different states. The state space is partitioned into different subsets of the power set of the set of states, with the information system as the partitioning agent, i.e., different information systems produce different partitions. Of course Demski's book has a wealth of ideas and constructs which cannot be explored here, but with this basic framework in mind, we note that results for algebraic systems are obtainable which differ from Demski's conclusions.

In effect the analysis presented here adds a level to the work of the information economists, exploring the possible derivations of, and treating the information systems used in, their formulations as information systems: these are representable as systems of first order formulas and consequently are amenable to analysis as structures in model theory. Whereas the information economics approach formalizes information systems as collections of functions from the states to the signals, our approach imposes additional constraints on the production of the signals themselves by explicitly considering the language used to express the functional formulation of the information systems. These additional constraints will be of consequence when the information system is represented as a proof-based system.

1.6. Location of the Research Justified

Returning to the article of Stephens et al. [1985] which was discussed in 1.4, we note their description of accounting interfaces. If this description is reduced to an individual firm, then the accounting system of the firm can be seen as a filter which captures certain data from the environment, to be processed and presented to decision makers. It is this filter, the specification of which data are captured, how they are processed and in what general form they are presented, which locate this book within the accounting process. In this location an accounting system is constructed as a machine which follows strict rules, namely the axioms, in converting inputs to outputs. This procedure is also strictly defined as an inference scheme, determining how occurrences of inputs combine within the rules to produce outputs and other secondary rules. These outputs are the derivates of the accounting system and may also arise from combining rules only from within the inference scheme. Thus we are dealing with the construction of a deterministic system.

In addition the book considers how to control accounting systems which operate under different rules. This requires building on the derivations of rule-based accounting systems. The control is achieved as follows. The derivations of an axiom system can be thought of as formal deductions from given premises. In this case, the formal deductions arise from the inference scheme and the premises are the axioms and derivations already deduced. The control of accounting systems under this methodology would then look at the differences in the sets of consequence of the separate proof-based accounting systems.

As mentioned previously, the Stephens et al. article describes accounting as an environment to accommodate accounting information systems (AIS) and decision maker flow. The link between the environment and the AIS and between the AIS and the decision maker are both areas of considerable research in accounting. The first link contains problem areas involving the recognition of economic events as transactions. Research has been concerned with when and whether an economic event such as a contingency should be captured by the system and thereafter reported to the decision maker. The crux of this problem is when an economic event should be interpreted as being probable. On the other hand, the link between the AIS and the decision maker involves the interpretation of whether and under what circumstances data presented by the AIS change decisions and thereby become information of some value to decision makers. Thus there are two general types of interpretation which occur between the AIS and its environment and the AIS and the decision makers. However, in the context developed in this book a third and more formal approach to interpretation is employed. This third type occurs entirely within the AIS and is specifically related to the axiomatization of the system. Within axiomatized systems there is a formal logical interpretation, indeed an interpretation function, between the syntactic components of the system and their semantic interpretations.

1.7. Accounting and Formal Languages

The axioms and derivations of a formal system are strings of symbols called sentences or formulas. At the syntactic level these strings are manipulated by the inference scheme in a purely formal way, without regard to the meanings which may be attached to the original or deduced sentences. The syntactic level therefore is merely concerned with which sentences can be produced by following the inference scheme. So the only way that a sentence is in essence "meaningless" in syntax is if it is not derivable from the axioms via a sequence of inferences. A logical interpretation is a formal map from the syntactic level to the semantic level which provides meaning or a translation of the combination of symbols in the sentences.

As an example, consider the standard rule of inference *modus ponens*. According to this rule, if there are two sentences $x \rightarrow P(x)$ and x, then $P(x)$ is derivable. Notice that no meaning is attached to the symbols x, $\rightarrow$ or $P(x)$, so that modus ponens is a strictly syntactic construction. Now suppose the interpretation function maps x to "cash", $\rightarrow$ to "implies" and $P(x)$ to "x is a current asset". Then at the semantic level the interpretation of this instance of modus ponens is that "cash implies cash is a current asset", so that "cash is a current asset" is derivable.

The syntactic rules are akin to the grammatical rules of a natural language. In natural languages the meaning of a sentence is based on an interpretation of its form. This form is regulated by distinguishing which sentences are grammatical. However, the distinction between syntax and semantics in natural language is often a hazy one because the grammatical rules are not specified in advance, but have been deduced from the structure of the language by linguists. Therefore it may be impossible to describe accurately the syntax of a language by a finite number of rules. For example, a basic sentence form in English is subject-verb-object. This rule works well for "sensible" sentences such as "The computer ran the program." Unfortunately, without further rules of grammatical construction, a naive foreign speaker might deduce the following sentence from the rule: "The computer walked the program." The purpose of indicating this type of problem in a natural language is to point out the close relationship of both syntax and semantics to the interpretation of meaning in these languages. It appears as if the human mind attends to both syntax and semantics simultaneously through learned patterns when constructing the meaning of natural language sentences.

Another type of interpretation known as hermeneutics has been developed in the naturalistic research methodology. By using this methodology the researcher attempts to interpret the world as a text in order to understand the meanings which the actors in the study attach to objects, to themselves and others, and to actions. Here the objective is similar to reducing the semantic context of the world to a somewhat less complex and perhaps hidden syntactic component. The syntactic component in hermeneutics is seen to be dynamic, with the actors and their environment constantly interacting to reconstitute meaning and form. This technique is essentially a meta-analysis of sentences which not only looks at sentences in their own contexts, but also across the contexts of different actors and environments.

In order to relate accounting to the concepts of syntax and semantics, it should be remembered that these concepts are used in different ways in various types of analysis. In the case of natural language, the syntactic component of accounting is the systems of rules, such as the mechanics of double entry bookkeeping, statements of auditing standards, Financial Accounting Standards Board (FASB) and International Accounting Standards Board publications, and Security and Exchange Commission rulings which affect transactions and manipulations of transactions, including disclosure. As with all natural languages, the syntactic and semantic components lie very close to one another when accounting is viewed as a natural language. For example, take a common occurrence when beginning students are introduced to accounting for merchandizing firms. A typical error is for the student to debit inventory and credit accounts payable when merchandize is purchased on account, instead of debiting purchases. This may happen because the student is confused about the semantic meaning of the problem of costs of goods sold, as against the meaning of accounting for inventories.

A further phenomenon which occurs when accounting is viewed as a natural language is that the interpretational component becomes closely intertwined with both the syntactic and the semantic components. The conceptual framework and the FASB statements which refine previous interpretations in order to standardize interpretation of economic events indicate the closeness of this relationship. For example, FASB statement number 1 is an attempt to standardize the interpretation of what constitutes an operating lease, as opposed to a capital lease for both the lessee and the lessor. This is similar in form to the hermeneutic concept of interpretation acting as the meta-rule intermediating between the actors and their environment, in this case certified public accountants, their clients, the FASB members and the accounting environment.

The explanation of the interrelationships between form, meaning and interpretation was the original inspiration to the formulation of formal logics and proof-based systems. The ancient Greeks were concerned with problems of valid arguments, proceeding from the development of schools of rhetoric. At the time work was concentrated on developing techniques for identifying correct inferences and exposing fallacious ones. One of the arguments which arose was between Diodorus Cronus and his pupil Philo of Megara. The argument revolved around the correct interpretation of the rule of inference modus ponens, which is also known as a conditional statement. If the conditional statement is formulated as "$a \rightarrow b$", then a is termed the antecedent and b the consequent. Diodorus and Philo differed as to what would be the conclusion if the antecedent were false. Diodorus took the position that a false antecedent negated the conditional, so that the statement is false. Thus the statement "If the FASB is a governmental agency, then this book is deposited" is false in Diodorus' system, since the FASB is not a governmental agency. Philo took the opposite view, arguing that the only case where the conditional is false is when the antecedent is true and the consequent is false.

Philo's reasoning is important because his position became the standard one in formal logic. Under his interpretation, $a \rightarrow b$, which semantically might be read as "a implies b" or "if a, then b", is logically equivalent to "not a or b". In this case, the "or" is interpreted as inclusive, meaning that "not a is true" or "b is true" or "both are true". (In the case of an exclusive or, the last case is disallowed.) Under this interpretation, treating the sentence "If the FASB is a governmental agency, then this book is deposited" is equivalent to "either the FASB is not a governmental agency or this book is deposited", which is true since the FASB is not currently a governmental agency. Notice that the second clause "this book is deposited" can be either true or false and the entire statement remains true as long as the FASB remains independent. As such, the Philonian interpretation of false antecedents is often referred to as the case of trivial truth of the conditional.

Whatever justification there may be for the specific interpretations that have been given to inference schemes, and there are many equivalences between rules of inference schemes as well, the point is that the construction of formal systems requires the specification of exact syntactic rules and specific interpretive mappings to semantic meaning. In addition, even after the specification of the formal inference scheme, it may be possible to reduce the number of allowed inferences by eliminating inferences which are logically equivalent to one another. For example, many of the inferences allowed in formal logics currently used in philosophical and linguistic texts on the subject were developed in the Middle Ages by the scholastics in order to match natural language inferences used in disputation and rhetoric. One such rule of inference is *modus tollens*, a type of negated modus ponens. With modus tollens the conclusion "not a" is deduced from "$a \rightarrow b$" and "not b".

This rule is equivalent to modus ponens, as can be seen when the conditional is translated into the form "not a" or "b"'. In the case of modus ponens, given "not b" along with the translated conditional, the only case where "not a or b" is true is when a is false, since then "not a" is true. This characterizes an important fact about formal inferences, they preserve truth. This means that if the premises of the inference, here "$a \rightarrow b$" and "not b", are true, then the conclusion must be true as well for the inference to be valid. The disadvantage of eliminating equivalent inferences is that it moves the logical system, as represented by the sentences, further away from natural language. This is true because the natural language inferences are translated into a reduced set of inferences, which removes some of the variety from the corresponding formal language. The variety lost does not entail a loss of content however, since the reduced system is logically equivalent to the system with the larger set of inferences. The reduced system does possess syntactic advantages however, since the number of rules has been reduced. This allows for simpler analysis at the syntactic level. In fact, many mathematical logic systems only include modus ponens in their inference schemes and these are almost always equivalent to systems which allow a greater number of inference types.

In this work we follow the formal systems of the mathematicians, rather than the philosophers and linguists, because the reduction in the number of inference rules reduces the complexity of specifying the consequences for computation. In order to prepare for the formal analysis which follows, some of the major concepts of the syntax and semantics of formal proof-based systems will now be introduced.

1.8. Proof-Based Systems

In order to formalize a language, there must be a specification of the signs and symbols of the formal language, as well as a specification of the permissible manipulations of the symbols. First an alphabet for the formal language is needed. The alphabet is divided into six disjoint subsets, the first of which are constants. Constants are symbols which have a single value such as 0 or 1. The second subset of the alphabet consists of variables, which can take on a range of values. Constants and variables are called atomic terms. The third subset consists of operations or functions. Each function has a specified degree $1, 2, 3, \ldots$; their values are called terms. Multiplication is an example of a function of degree 2 since it has two arguments. Functions map elements in their domains to elements in their codomains. They must be well-defined, meaning that each element in their domain is mapped to a unique element in the codomain. If f is a function of degree i and $t_1, t_2, \ldots, t_i$ are terms, then $f(t_1, t_2, \ldots, t_i)$ is also a term, although not an atomic term.

The next subset of the alphabet consists of predicates, which also have a specified degree. In effect a predicate makes a statement about its arguments. It does this because it is a defined subset of the domain of discourse or universe of the formal language. The universe contains all of the object-meanings which are allowed in the language. For example, if the universe consists of all of the accounts in an accounting system and a predicate $P(a)$ of degree 1 is defined to be "a is an asset", then $P(a)$ will be true only if a represents an asset account. This defines a mapping in which $P(a)$ is sent to "true" (or 1) in only those cases where a is in the subset P; otherwise $P(a)$ is mapped to "false" (or 0). This mapping is called the characteristic function of the predicate. If P is a predicate of degree i and $t_1, t_2, \ldots, t_i$ are terms, then $P(t_1, t_2, \ldots, t_i)$ is an atomic formula. Notice that it is possible to represent functions of degree i by predicates of degree $i+1$ by merely adding the codomain of the function as the $(i+1)$th object of the subset defined by the predicate. For example, the binary degree function of addition translates into a tertiary predicate in which $\langle 2, 5, 7 \rangle$ and $\langle 3, 8, 11 \rangle$ would be included in the subset defined by the addition predicate. In general this predicate would consist of the ordered triplets $\langle x, y, z \rangle$ such that $x + y = z$.

The fifth subset of the alphabet consists of logical symbols, which are divided into connectives and quantifiers. The connectives are $\rightarrow$ (implication), $\vee$ ("or" = disjunction), $\wedge$ ("and" = conjunction), $\neg$ ("not" = negation) and $\leftrightarrow$ (if and only if or logical equivalence). The quantifiers are $\exists$ (there exists) and $\forall$ (for all). The two quantifiers are also called the existential and the universal quantifiers respectively. If F and G are formulas, then the following are also formulas:

$$(F) \rightarrow (G), \ (F) \vee (G), \ (F) \wedge (G), \ \neg(F), \ (F) \leftrightarrow (G)$$

and

$$\exists(x)(F), \ \ \forall(x)(F),$$

where in the last two formulas x is a variable. The final subset contains punctuation marks, of which only left and right parentheses and occasional commas are used here.

The rules for forming terms and formulas provide the ability to recognize well-formed formulas in the language. A formula is well-formed if and only if it is built up from constants and variables by repeated application of the rules for forming terms and atomic formulas. In addition a formula in which all the variables are bound to quantifiers is called a sentence. A variable is bound if it occurs in a formula F and in the quantification of that formula, i.e., x is bound in F by the quantifications $\exists(x)(F)$ or $\forall(x)(F)$. A variable which is not bound is considered free.

Next the syntax and semantics of a formal language are constructed as follows. Both concepts are founded on the idea of the truth of formulas, sentences and inferences. Each logical symbol in the alphabet has a corresponding truth table associated with it. In the case of implication, the formula is false only when its antecedent is true and its consequent is false. Likewise, in the case of the inclusive or, the formula is false only when both arguments are false. For a conjunction, its truth value is true only when both arguments are true; in all other cases it is false. Negation takes only one argument and is true if its argument is false and false if its argument is true. Logical equivalence is true if and only if either both arguments are true or both are false.

The quantifiers "for all" and "there exists" are true in the following cases. "For all $x, F(x)$" is true only when every symbol of the alphabet which can be substituted for x in the formula leads to the formula being true. For "there exists $x, \ F(x)$", the formula is true if at least one symbol can be substituted for x leading to a true formula. The assignment of truth values for a complicated formula begins at the lowest level of atomic terms and atomic formulas and proceeds to higher levels, in the same manner as the term or formula was created in its definition.

It was mentioned earlier that in order for an inference to be valid, it must preserve truth. This means that it is not valid to deduce a false conclusion from true premises. The notion of validity is a syntactic one because it involves the construction of formulas through the application of the rules of inference. Given a set of axiom formulas, the formulas which can be validly constructed from the axioms by repeated inferences are called the consequences of the formulas and these are said to be deducible or derivable from the axiom formulas.

In terms of the semantic component of a formal language, all formulas that are true in the language are said to be provable in the language if the language is complete. Completeness is a semantic concept because it requires that if the meaning of some formula is true in the sense of the universe of the language, then that formula must be provable. The specific derivation of the formula does not have to be given however. Another general concept of formal languages is consistency. Consistency means that if a formula is derivable in the language, then its negation is not derivable. This is an important technical detail since, if both a formula and its negation can be proved in the language, then any formula in the language can be proved as well, a situation which certainly adds nothing to the sum of human knowledge.

1.9. The Scope of the Present Work

After this extended discussion of methodologies in accounting, the final section describes the scope of this book and what the authors believe is accomplished therein. The purpose of the book is to demonstrate how and under what conditions a basic accounting system can be reduced to a formal proof-based language. When this is accomplished, a method for controlling such systems through their derivations is established which is significantly stronger than methodologies used currently in accounting. The exposition in Chapters 2 through 9 employs definitions, propositions and proofs to formalize the system. The definitions are intended to represent terms,

concepts or constructions currently in use and are carefully stated
in order to avoid confusion as to the precise meaning assigned to
them in the book. Propositions are used to state results which follow logically from the definitions and are in all cases accompanied
by complete proofs. These proofs are meant to demonstrate the
correctness of the propositions and to illustrate the techniques used
in the algebraic and logical analyses.

Since the location of the research is the accounting system after
an economic event has occurred and been quantified, but before
the output of the system has been used by decision makers, the
method concentrates on the manipulation and processing of inputs
to outputs. These procedures are reduced to a purely algebraic
system which is capable of receiving transaction data, processing
the data and generating information in the form of summaries of
various types.

What happens when the accounting system is reduced to an algebraic system is that the entire range of speech is circumscribed. This
means that all sentences or ideas are known to be true, false or outside the particular system. Accounting systems can be thought of as
possessing different dialects, some quite similar, others nearly distinct. The control of accounting systems then takes on the quality of
distinguishing very precisely how the systems differ, i.e., which have
larger vocabularies and which are richer in expressiveness. From a
practical viewpoint this allows accountants as designers to match
the expressive power of particular systems to user needs for more
or less expressive languages. In addition the methodology can provide a means by which to identify situations of data or information
asymmetry and can therefore act as an indirect guide to action.

It cannot be claimed that this reduction is unique, for there are
many different opinions about what constitutes an accounting system and consequently many ways to construct a formal system. The
intention here is to provide a method which mirrors a specified basic
accounting system and which is reasonably comprehensible. It is not
the intention of the book to provide a blueprint for an accounting
system which could be programmed and used in practice. Rather
the concern is to allow the system to recognize and act on the transaction data itself. The base level justification is to develop a full
formal language for a particular aspect of accounting instead of assuming that such a grammar could exist and proceeding with partial

constructions or a higher level analysis. The success of this basic stage of proof-based research in accounting will furnish researchers with a secure, well established base for future investigations.

Algebraic concepts employed

It is time to be specific about the algebraic concepts that have proved useful in the analysis. There are four principal structures which are used repeatedly and which appear well suited to application in accounting, namely:

- balance vector;

- directed graph (or digraph);

- automaton;

- monoid.

These structures will be familiar to most algebraists. A few words will be given to elucidate their meaning and to justify the claim of utility in accounting theory.

A *balance vector* is a column vector or column matrix the sum of whose entries equals zero. In this case the relevance to accounting will be obvious: the zero sum reflects the fundamental property of any accounting system that it must always be in balance. Mathematicians will immediately recognize that balance vectors form a structure with known algebraic properties; they form a submodule or hyperplane. Balance vectors are able to represent the state of an accounting system at any instant. They are also capable of encoding the transactions that are applied to the system. There is an important comment to be made regarding signs: for the entries of a balance vector can be positive or negative. The great advantage of using positive and negative signs is that the signs take care of questions of credit or debit automatically; for example, a credit balance has a positive sign and a debit balance a negative one. The theory of balance vectors is developed in Chapters 2 and 3, where their application to accounting is clearly laid out.

The second useful algebraic notion is that of a *directed graph*. This is best thought of geometrically, although its definition is entirely algebraic. The digraph consists of *vertices*. i.e., points in the plane, and *edges*, or lines with a direction, joining certain vertices.

The vertices represent accounts and the edges indicate where there are flows of value within the system. Thus a digraph gives a picture of how value can flow around an accounting system. While in general different accounting systems might have the same digraph, for certain special types the digraph determines the system up to equivalence.

The third concept, that of an *automaton*, is frequently used in information science as a theoretical model of a computer. The automaton is at any instant in a certain state; it reads a symbol on an input tape, goes to a another state and then writes a symbol on an output tape. The applicability to accounting is clear: the states of the accounting system are the balance vectors, the inputs are the transactions and the outputs are the new balance vectors. This simple picture can be made more complex in order to represent further actions of an accounting system, as is expounded in detail in Chapters 6 and 9.

The final concept of a *monoid* is the most abstract. Every automaton has an associated monoid, which is an algebraic structure with a means of combining its elements subject to suitable rules. An input to the automaton produces a change in the state of the automaton and thus determines a function from states to states. The functions on the set of states form a monoid for which the operation is functional composition; the associated functions generate a submonoid of this monoid. Despite their abstraction, monoids provide useful ways of characterizing accounting systems with special properties, as is shown in Chapter 7.

With the aid of the concept of a balance vector, the definition of an abstract accounting system is laid out in Chapter 4 and its properties are expounded, with numerous accompanying examples. Relations between different accounting systems are considered in Chapter 5 by using standard constructions from algebra, namely quotient systems and homomorphisms. The latter are functions between different accounting systems that relate their structures.

An important topic in algebra is the possible existence of algorithms to perform certain computations or to make decisions: what is at stake here is the question of what can and cannot be computed, in principle at least. For example, is it possible to write a program which is able to test the final balance vector of an accounting system

and decide if there have been any irregularities during the accounting period? The importance of the question is evident. Chapter 8
contains a full discussion of what one can expect to be able to decide
or compute in an accounting system.

In Chapter 9 all the strands come together to form our final
model of an accounting system. In this there are ten parameters, so
the model is referred to as the *10-tuple model*. It has the capability
to scan and process incoming transactions, keep track of balances,
generate reports on the system, control access by individuals to the
system, and keep track of frequency of application of transactions. It
is also able to test final balances. Our main conclusion in this book is
that the 10-tuple model goes a long way towards representing what
is actually going on during the operation of an accounting system.

The final Chapter 10 is intended as a corrective after the many
mathematical considerations of this work. It presents a detailed
example of a small company engaged in trade and it exhibits the
accounting system in the form of a 10-tuple model. The aim of
the example is, of course, to help make the case for the relevance
of the model to accounting practice and to justify the claim that
all connection with reality has not been been eroded through the
process of abstraction.

Chapter Two

Balance Vectors

In this chapter we begin the task of assembling the various components of our basic algebraic model of an accounting system. The first concern is to provide a means of describing the state of an accounting system at any instant. Now this is most naturally accomplished by listing the "values" of the various accounts in the system. So the first step must be to identify an algebraic structure to which the account values will belong. The structure chosen must be sufficiently rich to accommodate the operations that one would expect to apply to the accounting system. It emerges from the discussion below that *ordered integral domains* are the natural candidates.

Once the domain of account values has been settled, the list of account values can be conveniently displayed as a column vector. Such column vectors will have the special property that the sum of all their entries is zero, which reflects the requirement that the accounting system should always be in balance. Column vectors with entry sum equal to zero are called *balance vectors* and they form the foundation of our theory. For this reason the chapter focuses on *balance vectors over ordered domains* and associated mathematical structures.

2.1. The Values of an Account

Our first task is to analyze the precise requirements demanded of the values of an account. Of course in practice such values would likely be in a currency such as dollars or euros, although numbers of any items of value would also be a possibility; thus integers or real numbers would be appropriate for account values. But the question to be addressed is: what formal properties should account values actually have?

One obvious requirement is the ability to add and subtract values. This is clearly essential in any accounting system. From the purely accounting point of view there is no reason to be able to multiply account values. On the other hand, it turns out that there are good mathematical reasons for the introduction of a multiplication operation: for it leads to increased richness of mathematical structure by allowing the use of modules and hence the methods of linear algebra. However, it should be stressed that multiplication is introduced as a mathematical device and it does not carry significance for accounting.

Naturally the addition, subtraction and multiplication should satisfy reasonable rules, by which we will mean the standard rules of arithmetic. Now it is time to make all of this precise.

Let there be given a set R together with two binary operations on R called *addition* and *multiplication*, denoted in the usual way, such that the following rules hold for all elements a, b, c of R:

1. $(a + b) + c = a + (b + c)$ (associative law);

2. $a + b = b + a$ (commutative law);

3. R contains a *zero element*, written 0_R or 0, such that $a + 0 = a$ for all a in R;

4. each element a of R has a *negative* $-a$ of R, with the property that $a + (-a) = 0$;

5. $(ab)c = a(bc)$ (associative law);

6. $ab = ba$ (commutative law);

7. $a(b + c) = ab + ac$ (distributive law);

8. R contains an *identity element*, written 1_R or 1, such that $a1 = a$ for all a in A.

The first four of these requirements assert that R is an algebraic structure called an *abelian group* (after N. Abel). With the additional properties (5) through (8) R becomes a *commutative ring with identity*. Note that subtraction in R can be defined by the rule $a - b = a + (-b)$. Thus a commutative ring with identity is an algebraic structure in which one can add, subtract and multiply subject to the usual rules of arithmetic. Notice however that division is not permitted.

The values of the accounts in an accounting system will be elements of a commutative ring with identity R. However the structure of R is still not rich enough. For it is an essential feature of an accounting system that the value of an account can be regarded as positive or negative (or zero, of course): here the standard convention is that the value of an account representing an asset should normally be positive, while the value of a liability account should be negative: other accounts such as profit or loss could have positive or negative values. In any event we recognize that the ring R must admit the concept of "positive" and "negative" elements. This calls for the introduction of an order relation on the domain R.

A commutative ring with identity R is said to be *linearly ordered* if there is a non-empty subset P of R not containing 0, called the set of *positive elements*, such that the following conditions are satisfied:

9. if $a, b \in P$, then $a + b \in P$ and $ab \in P$;

10. for each $a \in R$, one of the following holds: $a \in P$, $a = 0$, $-a \in P$.

The actual concept of a *linear order* arises when one defines

$$a < b$$

to mean that $b - a \in P$. The *negative elements* of R are the elements of the set $R \backslash (P \cup \{0\})$. On the basis of (9) and (10) it can be shown that the following holds:

11. for any $a, b \in R$, exactly one of the statements $a < b$, $a = b$, $b < a$ holds.

There is another important and easily deduced consequence of (9) and (10):

12. if $ab = 0$ with $a, b \in R$, then $a = 0$ or $b = 0$.

A commutative ring with identity which satisfies (11) is called an *integral domain*, or simply a *domain*. Thus the ordered commutative rings with identity are exactly the *ordered domains*. The most obvious examples of ordered domains are

$$\mathbb{Z}, \quad \mathbb{Q}, \quad \mathbb{R},$$

the sets of integers, rational numbers, real numbers respectively, where the standard arithmetic operations of addition and multiplication, and the usual meaning of "positive", are used. It is known that any ordered domain R has *characteristic zero*; this means that the equation $na = 0$, where $n \in \mathbb{Z}, a \in R$, implies that $n = 0$ or $a = 0_R$. This in turn shows that $\mathbb{Z}$ is always contained inside an ordered domain, so that it is the smallest possible candidate for R.

On the basis of the foregoing analysis, for accounting and mathematical reasons, we choose to make the values of accounts belong to an ordered domain. While more general algebraic structures might be envisaged, there is a convincing case that ordered domains provide the most natural realm for the values of the accounts in an accounting system.

2.2. The State of an Accounting System

Let R be an ordered domain, which will be the universal set for all account values, and let n be a positive integer, which will be the number of accounts in the accounting system. The state of the system at any instant can be described by listing the values of the accounts, which are assumed to be in some agreed order, in the form of an *n-column vector over R*,

$$\mathbf{v} = \begin{bmatrix} v_1 \\ v_2 \\ \vdots \\ v_n \end{bmatrix}.$$

Thus $v_i \in R$ is the value of the ith account. The set of all n-column vectors over R is denoted by

$$R^n.$$

Notice that R^1 and R are essentially identical. Of particular importance is the *zero vector*

$$\mathbf{0} = \begin{bmatrix} 0 \\ 0 \\ \vdots \\ 0 \end{bmatrix}.$$

There are two natural operations which may be applied to R^n and which are inherited from the ring R itself, namely addition and multiplication by elements of R. To specify these operations, let $\mathbf{u}$ and $\mathbf{v}$ belong to R^n and let $r \in R$. The *sum* $\mathbf{u} + \mathbf{v}$ is defined as in matrix algebra by adding corresponding entries,

$$\mathbf{u} + \mathbf{v} = \begin{bmatrix} u_1 + v_1 \\ u_2 + v_2 \\ \vdots \\ u_n + v_n \end{bmatrix},$$

while the scalar multiple $r\mathbf{v}$ is formed by multiplying each entry of $\mathbf{v}$ by r, again just as in matrix algebra,

$$r\mathbf{v} = \begin{bmatrix} rv_1 \\ rv_2 \\ \vdots \\ rv_n \end{bmatrix}.$$

On the basis of these familiar definitions, one can quickly verify that the operations of addition and scalar multiplication in R^n enjoy the following properties. Let $\mathbf{u}, \mathbf{v}, \mathbf{w} \in R^n$ and $r, s \in R$:

1. $(\mathbf{u} + \mathbf{v}) + \mathbf{w} = \mathbf{u} + (\mathbf{v} + \mathbf{w})$;

2. $\mathbf{u} + \mathbf{v} = \mathbf{v} + \mathbf{u}$;

3. $\mathbf{v} + \mathbf{0} = \mathbf{v}$;

4. $\mathbf{v} + (-\mathbf{v}) = \mathbf{0}$;

5. $r(\mathbf{u} + \mathbf{v}) = r\mathbf{u} + r\mathbf{v}$;

6. $(r + s)\mathbf{u} = r\mathbf{u} + s\mathbf{u}$;

7. $(rs)\mathbf{v} = r(s\mathbf{v})$;

8. $1_R\mathbf{v} = \mathbf{v}$.

Here $-\mathbf{v}$, the *negative* of $\mathbf{v}$, arises on changing the sign of each entry of $\mathbf{v}$; clearly $-\mathbf{v} = (-1_R)\mathbf{v}$.

These properties demonstrate that the set R^n has a recognizable algebraic structure. Indeed properties (1) – (4) assert that R^n is an abelian group, while the additional properties (5) through (8) make R^n into a (*left*) *R-module*. Thus in an R-module one can add and subtract, and also multiply by elements of the ordered domain R, all subject to the rules above.

The free R-module R^n

It turns out that R^n is a particular type of R-module called a *free R-module*. To see what is special about it, consider the so-called *elementary* column vectors $\mathbf{e}(1), \mathbf{e}(2), \ldots, \mathbf{e}(n)$ where the ith entry of $\mathbf{e}(i)$ is $1 = 1_R$ and all other entries are 0. Thus

$$\mathbf{e}(1) = \begin{bmatrix} 1 \\ 0 \\ \vdots \\ 0 \end{bmatrix}, \quad \mathbf{e}(2) = \begin{bmatrix} 0 \\ 1 \\ 0 \\ \vdots \\ 0 \end{bmatrix}, \ldots, \quad \mathbf{e}(n) = \begin{bmatrix} 0 \\ 0 \\ \vdots \\ 0 \\ 1 \end{bmatrix}.$$

Now an arbitrary vector $\mathbf{v}$ in R^n is expressible in terms of these elementary vectors since

$$\mathbf{v} = v_1\,\mathbf{e}(1) + v_2\,\mathbf{e}(2) + \cdots + v_n\,\mathbf{e}(n),$$

i.e., $\mathbf{v}$ is an *R-linear combination* of $\mathbf{e}(1), \mathbf{e}(2), \ldots, \mathbf{e}(n)$. If this linear combination equals $\mathbf{0}$, then the equation shows that $\mathbf{v} = \mathbf{0}$ and so $v_1 = \cdots = v_n = 0$. Thus the only linear combination of $\mathbf{e}(1), \ldots, \mathbf{e}(n)$ that equals $\mathbf{0}$ is the one with all coefficients equal to 0. This means that $\mathbf{e}(1), \ldots, \mathbf{e}(n)$ are *linearly independent vectors*.

As in linear algebra, a subset S of R^n is called an R-*basis* of R^n if the elements of S are linearly independent and if each vector of R^n can be written as a linear combination of vectors in S. Moreover, this expression as a linear combination will be unique because of linear independence. Thus $\{\mathbf{e}(1), \ldots, \mathbf{e}(n)\}$ is an R-basis of R^n. An R-module which has an R-basis consisting of n elements is called a *free R-module of rank n*. (It is a known result from commutative ring theory that all bases of a free module have the same number of elements). Thus the previous discussion leads to:

(2.2.1). *The set of elementary vectors* $\{\mathbf{e}(1), \ldots, \mathbf{e}(n)\}$ *is an R-basis of* R^n, *so that* R^n *is a free R-module of rank n.*

Balance vectors in R^n

So far arbitrary column vectors over an ordered domain R have been considered. Now our aim is to use such vectors to express the

state of an accounting system by listing the balances of the various
accounts as the entries of the vector. However, the vectors to be
used must have the property that the sum of their entries is zero.
For it is an essential feature of the double entry accounting system
that it must always be in balance. To see what this entails, recall
that the accounts of a company generally fall into three categories:

1. *Asset accounts*, which represent anything owned by the company;

2. *Liability accounts*, which record what is owed by the company
   to external entities;

3. *Equity accounts* or a *profit and loss account*; these show what
   is owed by the company to the owners and also show the net
   assets of the company.

(There may also be temporary revenue and expense accounts). This
scheme is generally referred to as the *chart of accounts*.

It is a fundamental fact that a double entry accounting system
must always be in balance, a fact which is implied by *the accounting
equation*
$$A - L = E$$
where A, L and E are respectively the totals of all amounts in asset
accounts, liability accounts and equity or profit and loss accounts.
In general the asset accounts will have positive balances and liability
accounts negative balances. The accounting equation may also be
written in the form
$$A - L - E = 0,$$
an equation which shows that the total equity, or net assets, of the
company should have a negative sign, at least if the company is
making a profit. Indeed the negative sign is to be expected since
the amount E is owed by the company to the owners.

The second equation causes us to focus on column vectors in R^n
whose entry sum is 0, the so-called *balance vectors over R*. For
at any point in time the state of the company's accounting system
is described by the column vector whose entries are the account
balances *with the appropriate signs*, in short by a balance vector.

We proceed now to study the properties of the subset of balance vectors in R^n, where R is an ordered domain. Keep in mind that R and R^1 are considered to be identical. First let us consider the function

$$\sigma : R^n \to R$$

which sums the entries of a column vector $\mathbf{v}$ in R^n, i.e.

$$\sigma(\mathbf{v}) = \sum_{i=1}^{n} v_i.$$

It is simple to check that σ has the following properties:

$$\sigma(\mathbf{v} + \mathbf{w}) = \sigma(\mathbf{v}) + \sigma(\mathbf{w}), \quad \sigma(r\mathbf{v}) = r\sigma(\mathbf{v})$$

for all $\mathbf{v}, \mathbf{w} \in R^n$ and $r \in R$. A function between two R-modules with these properties is called an *R-module homomorphism*.

The reason for introducing the function σ is that the vectors $\mathbf{v}$ which satisfy $\sigma(\mathbf{v}) = 0$ are exactly the balance vectors in R^n. Now module elements which are sent to zero by a module homomorphism σ form a subset called the *kernel*, written

$$\mathrm{Ker}(\sigma).$$

It is easily checked that the kernel is itself a module, which is of course contained in the domain of the homomorphism. Thus the kernel of a module homomorphism is a *submodule* of the domain.

Returning to the particular homomorphism σ, we conclude that its kernel, i.e. the set of balance vectors, is a submodule of R^n. We shall write

$$\mathrm{Bal}_n(R)$$

for the set of all balance vectors in R^n, so that $\mathrm{Ker}(\sigma) = \mathrm{Bal}_n(R)$ is a submodule of R^n, which will be called the *balance module* of degree n over R.

Examples of balance vectors

(i) There is just one balance vector in $R^1 = R$, namely the zero vector $\mathbf{0}$. A typical balance vector in R^2 has the form

$$\begin{bmatrix} a \\ -a \end{bmatrix}, \quad (a \in R),$$

while a general balance vector in R^n can be written as

$$\begin{bmatrix} a_1 \\ a_2 \\ \vdots \\ a_{n-1} \\ -a_1 - a_2 - \cdots - a_{n-1} \end{bmatrix}$$

where $a_i \in R$.

(ii) An especially important type of balance vector occurs when there are just two non-zero entries, one of which will of course have to be the negative of the other. Such vectors are called *simple balance vectors*. As an example of a simple balance vector, consider

$$\mathbf{e}(i,j), \quad i \neq j,$$

which is the vector in R^n, $(n \geq 2)$, whose ith entry is 1 and jth entry is -1, with all other entries 0. Then $\mathbf{e}(i,j)$ is a simple balance vector in R^n. For example, when $n = 4$,

$$\mathbf{e}(2,3) = \begin{bmatrix} 0 \\ 1 \\ -1 \\ 0 \end{bmatrix} \quad \text{and} \quad \mathbf{e}(3,1) = \begin{bmatrix} -1 \\ 0 \\ 1 \\ 0 \end{bmatrix}.$$

The $\mathbf{e}(i,j)$ are called *elementary balance vectors*: evidently every simple balance vector is a scalar multiple of an elementary balance vector. The number of elementary balance vectors in R^n is equal to

$$n(n-1),$$

for this is the number of ways of choosing two objects from a set of n in a definite order. In the following section it will be seen that the elementary balance vectors play a special role in the module $\mathrm{Bal}_n(R)$.

2.3. Properties of the Balance Module

Now that the balance vectors over an ordered domain R have been identified as the medium for expressing the state of an accounting system, we will take the opportunity to develop some of the mathematical properties of the module $\text{Bal}_n(R)$ of all balance vectors in R^n.

Our first observation is that $\text{Bal}_n(R)$, like R^n, is a free R-module, although it is of rank one less. Note that $\text{Bal}_1(R) = 0$, which is free of rank 0, so we can assume that $n > 1$. To prove the assertion it is necessary to produce an R-basis consisting of $n - 1$ vectors. This is done in the next result.

(2.3.1). *Let R be an ordered domain and let $n > 1$ be an integer. Then the elementary balance vectors $\mathbf{e}(1, 2), \mathbf{e}(2, 3), \ldots, \mathbf{e}(n - 1, n)$ constitute an R-basis of $\text{Bal}_n(R)$. Thus $\text{Bal}_n(R)$ is a free R-module of rank $n - 1$.*

Proof

In the first place $\mathbf{e}(1, 2), \ldots, \mathbf{e}(n - 1, n)$ are linearly independent. For if $r_1, \ldots, r_{n-1} \in R$, then

$$r_1\,\mathbf{e}(1, 2) + r_2\,\mathbf{e}(2, 3) + \cdots + r_{n-1}\,\mathbf{e}(n - 1, n) = \begin{bmatrix} r_1 \\ r_2 - r_1 \\ r_3 - r_2 \\ \vdots \\ r_{n-1} - r_{n-2} \\ -r_{n-1} \end{bmatrix},$$

and the only way this can equal $\mathbf{0}$ is if $r_1 = r_2 = \cdots = r_{n-1} = 0$.

It remains to prove that an arbitrary balance vector $\mathbf{b}$ is expressible as a linear combination of $\mathbf{e}(1, 2), \ldots, \mathbf{e}(n - 1, n)$. Write

$$\mathbf{b} = \begin{bmatrix} b_1 \\ b_2 \\ \vdots \\ b_{n-1} \\ -b_1 - b_2 - \cdots - b_{n-1} \end{bmatrix}$$

and define v_i to be $b_1 + b_2 + \cdots + b_i$, where $1 \leq i \leq n - 1$. Then

$v_1 \, \mathbf{e}(1,2) + v_2 \, \mathbf{e}(2,3) + \cdots + v_{n-1} \, \mathbf{e}(n-1,n)$ equals

$$
\begin{bmatrix} b_1 \\ -b_1 \\ 0 \\ \vdots \\ 0 \end{bmatrix}
+
\begin{bmatrix} 0 \\ b_1 + b_2 \\ -b_1 - b_2 \\ 0 \\ \vdots \\ 0 \end{bmatrix}
+ \cdots +
\begin{bmatrix} 0 \\ 0 \\ 0 \\ \vdots \\ 0 \\ b_1 + b_2 + \cdots + b_{n-1} \\ -b_1 - b_2 - \cdots - b_{n-1} \end{bmatrix},
$$

which, by a simple computation with column vectors, reduces to

$$
\begin{bmatrix} b_1 \\ b_2 \\ b_3 \\ \vdots \\ b_{n-1} \\ -b_1 - \cdots - b_{n-1} \end{bmatrix}.
$$

Hence $\mathbf{b} = v_1 \, \mathbf{e}(1,2) + v_2 \, \mathbf{e}(2,3) + \cdots + v_{n-1} \, \mathbf{e}(n-1,n)$. $\square$

One can think of 2.3.1 as saying that $\mathrm{Bal}_n(R)$ is rather similar to the module R^{n-1}. The next result shows how $\mathrm{Bal}_n(R)$ is situated within R^n.

Let $\mathbf{u}$ be a vector in R^n whose entry sum equals 1, i.e.

$$
\sum_{i=1}^{n} u_i = 1.
$$

There are, of course, many such vectors in R^n: for example, one could take one entry of $\mathbf{u}$ to be 1 and all the others to be 0. Denote by

$$
R\mathbf{u}
$$

the set of all multiples of $\mathbf{u}$ by elements of R. Then $R\mathbf{u}$ is a submodule of R^n and in fact it is a free R-module of rank 1 since $\{\mathbf{u}\}$ constitutes an R-basis for it.

Now choose any $\mathbf{v}$ from R^n and put $r = \sum_{i=1}^{n} v_i$. Then the vector $\mathbf{v} - r\mathbf{u}$ has entry sum

$$
\sum_{i=1}^{n} (v_i - r u_i) = \sum_{i=1}^{n} v_i - r \sum_{i=1}^{n} u_i = r - r1 = 0
$$

since

$$\sum_{i=1}^{n} v_i = r \quad \text{and} \quad \sum_{i=1}^{n} u_i = 1.$$

Thus $\mathbf{v} - r\mathbf{u}$ is a balance vector. Since $\mathbf{v} = (\mathbf{v} - r\mathbf{u}) + r\mathbf{u}$, it follows that every vector in R^n is the sum of a vector in $\text{Bal}_n(R)$ and a vector in $R\mathbf{u}$; this is expressed by the equation

$$R^n = \text{Bal}_n(R) + R\mathbf{u}.$$

Next, if $\mathbf{v} \in \text{Bal}_n(R) \cap R\mathbf{u}$, then $\mathbf{v} = r\mathbf{u}$ for some $r \in R$. Also

$$0 = \sum_{i=1}^{n} v_i = \sum_{i=1}^{n} r u_i = r \sum_{i=1}^{n} u_i = r.$$

Thus $\mathbf{v} = \mathbf{0}$ and therefore

$$\text{Bal}_n(R) \cap R\mathbf{u} = 0,$$

the zero submodule.

The two statements combine to say that R^n is the *direct sum* of the submodules $\text{Bal}_n(R)$ and $R\mathbf{u}$, in symbols

$$R^n = \text{Bal}_n(R) \oplus R\mathbf{u}.$$

This conclusion is stated formally in the next result.

(2.3.2). *The R-module R^n is the direct sum of the submodules $\text{Bal}_n(R)$ and $R\mathbf{u}$, i.e., $R^n = \text{Bal}_n(R) \oplus R\mathbf{u}$, where $\mathbf{u}$ is any vector such that $\sum_{i=1}^{n} u_i = 1$.*

Direct sums of more than two submodules can be defined by iteration and will occasionally be used in the sequel.

Balance vectors and permutations

From the algebraic point of view a natural way to generate balance vectors is to take an arbitrary vector in R^n and subtract from it a vector obtained by permuting its entries. The resulting vector will always be a balance vector. For example, when $n = 4$, one might form

$$\begin{bmatrix} v_1 \\ v_2 \\ v_3 \\ v_4 \end{bmatrix} - \begin{bmatrix} v_1 \\ v_4 \\ v_3 \\ v_2 \end{bmatrix} = \begin{bmatrix} 0 \\ v_2 - v_4 \\ 0 \\ v_4 - v_2 \end{bmatrix},$$

which is a simple balance vector.

In general let π be a permutation of the integers $1, 2, \ldots, n$, i.e., π is a bijection from the set $\{1, 2, \ldots, n\}$ to itself. For each $\mathbf{v}$ in R^n form a new vector by using π to permute the entries of $\mathbf{v}$; this is the vector

$$\pi(\mathbf{v})$$

whose ith entry is $v_{\pi^{-1}(i)}$. Then $\mathbf{v} - \pi(\mathbf{v})$ is a balance vector since its has entry sum

$$\sum_{i=1}^{n}(v_i - v_{\pi^{-1}(i)}) = \sum_{i=1}^{n} v_i - \sum_{i=1}^{n} v_{\pi^{-1}(i)} = 0.$$

Hence $\mathbf{v} - \pi(\mathbf{v})$ belongs to $\mathrm{Bal}_n(R)$ and we have a function

$$\theta_\pi : R^n \to \mathrm{Bal}_n(R)$$

defined by the rule $\theta_\pi(\mathbf{v}) = \mathbf{v} - \pi(\mathbf{v})$.

There are some natural questions one can ask about the function θ_π. Is it a module homomorphism? If so, what are the kernel $\mathrm{Ker}(\theta_\pi)$ and the image $\mathrm{Im}(\theta_\pi)$? In particular, when is it surjective, i.e., when does every balance vector arise by applying the function θ_π to a suitable vector in R^n?

The next theorem answers these questions. First it is necessary to recall some basic facts about permutations. Any permutation π of $\{1, 2, \ldots, n\}$ is uniquely expressible up to order as a product, i.e., composite, of disjoint *cycles*, or cyclic permutations

$$\pi = \sigma_1 \circ \sigma_2 \circ \cdots \circ \sigma_k,$$

where each σ_i is a cycle of the form $(\ell_{i1}\ell_{i2}\ldots\ell_{ij_i})$ and the subsets $\{\ell_{i1}, \ell_{i2}, \ldots, \ell_{ij_i}\}$ of the set $\{1, 2, \ldots, n\}$ are disjoint. Recall that a cycle $(m_1, m_2, \ldots, m_r)$ sends m_1 to m_2, m_2 to m_3, $\ldots$, m_{k-1} to m_k and finally m_k to m_1, while other integers are fixed.

(2.3.3). *Let R be an ordered domain, n a positive integer and π a permutation of $1, 2, \ldots, n$. Assume that $\pi = \sigma_1 \circ \sigma_2 \circ \cdots \circ \sigma_k$ is the disjoint cycle decomposition of π. Then:*

1. *the function $\theta_\pi : R^n \to \mathrm{Bal}_n(R)$ is a homomorphism of R-modules;*

2. *the kernel of θ_π consists of those vectors $\mathbf{v}$ in R^n such that all entries coming from the same cycle σ_i are equal;*

3. *let $\sigma_i = (\ell_{i1}\ell_{i2}\ldots\ell_{ij_i})$ for $i = 1, 2, \ldots, k$. Then the image of θ_π has an R-basis consisting of all $e(\ell_{it}, \ell_{it+1})$, where $t = 1, 2, \ldots, j_i - 1, i = 1, 2, \ldots, k$.*

Proof

1. To prove this one simply has to verify that the requirements for a module homomorphism are satisfied. Notice first that $\pi(\mathbf{v} + \mathbf{w}) = \pi(\mathbf{v}) + \pi(\mathbf{w})$ and $\pi(r\mathbf{v}) = r\pi(\mathbf{v})$. Then

$$\begin{aligned}
\theta_\pi(\mathbf{v} + \mathbf{w}) &= (\mathbf{v} + \mathbf{w}) - \pi(\mathbf{v} + \mathbf{w}) \\
&= (\mathbf{v} - \pi(\mathbf{v})) + (\mathbf{w} - \pi(\mathbf{w})), \\
&= \theta_\pi(\mathbf{v}) + \theta_\pi(\mathbf{w})
\end{aligned}$$

and similarly

$$\begin{aligned}
\theta_\pi(r\mathbf{v}) = r\mathbf{v} - \pi(r\mathbf{v}) &= r(\mathbf{v} - \pi(\mathbf{v})) \\
&= r\theta_\pi(\mathbf{v}).
\end{aligned}$$

2. Let $\mathbf{v} \in R^n$. Then $\mathbf{v} \in \text{Ker}(\theta_\pi)$ if and only if $\theta_\pi(\mathbf{v}) = 0$, i.e. $\mathbf{v} = \pi(\mathbf{v})$. This says that $v_i = v_{\pi(i)}$ for all i, from which the statement follows at once.

3. This is harder to see. The crucial point to keep in mind is that π permutes the entries in each cycle $(\ell_{i1}\ell_{i2}\ldots\ell_{ij_i})$ according to the cyclic permutation σ_i. For any $\mathbf{v}$ in R^n we can write

$$\mathbf{v} = \mathbf{v}_1' + \mathbf{v}_2' + \cdots + \mathbf{v}_k'$$

where $\mathbf{v}_i'$ is the vector whose non-zero entries are the entries of $\mathbf{v}$ which correspond to components of the cycle σ_i. Then $\pi(\mathbf{v}_i') = \sigma_i(\mathbf{v}_i')$ and thus

$$\theta_\pi(\mathbf{v}) = \mathbf{v} - \pi(\mathbf{v}) = \sum_{i=1}^{k}(\mathbf{v}_i' - \pi(\mathbf{v}_i')) = \sum_{i=1}^{k}(\mathbf{v}_i' - \sigma_i(\mathbf{v}_i')) = \sum_{i=1}^{k}\theta_{\sigma_i}(\mathbf{v}_i').$$

Therefore it is enough to prove the statement for the cycle σ_i, i.e., we can assume that π is an n-cycle.

We may suppose without loss of generality that $\pi = (1 \ 2 \ \ldots \ n)$: then by 2.3.1 we need to show that $\text{Im}(\theta_\pi) = \text{Bal}_n(R)$. Choose any $\mathbf{u}$ in $\text{Bal}_n(R)$ and let its entries be

$$u_1, u_2, \ldots, u_{n-1}, -u_1 - u_2 - \cdots - u_{n-1}.$$

Define $\mathbf{v} \in R^n$ to be the vector with entries

$$u_1, \ u_1 + u_2, \ \ldots, \ u_1 + u_2 + \cdots + u_{n-1}, \ 0.$$

Then we have

$$\theta_\pi(\mathbf{v}) = \begin{bmatrix} u_1 \\ u_1 + u_2 \\ u_1 + u_2 + u_3 \\ \cdot \\ \cdot \\ u_1 + u_2 + \cdots + u_{n-1} \\ 0 \end{bmatrix} - \begin{bmatrix} 0 \\ u_1 \\ u_1 + u_2 \\ \cdot \\ \cdot \\ u_1 + u_2 + \cdots + u_{n-2} \\ u_1 + u_2 + \cdots + u_{n-1} \end{bmatrix},$$

which equals

$$\begin{bmatrix} u_1 \\ u_2 \\ u_3 \\ \cdot \\ \cdot \\ u_{n-1} \\ -u_1 - u_2 - \cdots - u_{n-1} \end{bmatrix} = \mathbf{u}.$$

Our conclusion is that when π is an n-cycle, $\mathrm{Im}(\theta_\pi)$ equals $\mathrm{Bal}_n(R)$. The statement in (3) now follows on applying this result to each of the cyclic permutations σ_i. $\square$

(2.3.4). *The module $\mathrm{Im}(\theta_\pi)$ is a free R-module of rank $n - k$, where k is the number of disjoint cycles in π.*

Proof
Let $j_1, j_2, \ldots, j_k$ be the lengths of the disjoint cycles of π. Then by 2.3.3 there is an isomorphism, i.e., a bijective homomorphism, from $\mathrm{Bal}_n(R)$ to

$$\mathrm{Bal}_{j_1}(R) \ \oplus \cdots \oplus \mathrm{Bal}_{j_k}(R),$$

which by 2.3.1 is a free R-module with rank

$$\sum_{i=1}^{k}(j_i - 1) = \left(\sum_{i=1}^{k} j_i\right) - k = n - k.$$

$\square$

From this it follows that $\text{Im}(\theta_\pi) = \text{Bal}_n(R)$ if and only if $n - k = n - 1$, i.e., $k = 1$; this is because $\text{Bal}_n(R)$ has rank $n - 1$. Thus we have:

(2.3.5). *The homomorphism θ_π is surjective if and only if π is an n-cycle.*

To illustrate the proof of 2.3.3 we present an example.

Example (2.3.1).

Let $n = 5$ and choose π to be the permutation $(1\ 2\ 3)(4\ 5)$. By 2.3.3 a general element of $\text{Im}(\theta_\pi)$ should have the form

$$\mathbf{u} = \begin{bmatrix} u_1 \\ u_2 \\ -u_1 - u_2 \\ u_3 \\ -u_3 \end{bmatrix}.$$

Following the method of the proof of 2.3.3, we form the vector

$$\mathbf{v} = \begin{bmatrix} u_1 \\ u_1 + u_2 \\ 0 \\ u_3 \\ 0 \end{bmatrix}.$$

Then

$$\theta_\pi(\mathbf{v}) = \mathbf{v} - \pi(\mathbf{v}) = \begin{bmatrix} u_1 \\ u_1 + u_2 \\ 0 \\ u_3 \\ 0 \end{bmatrix} - \begin{bmatrix} 0 \\ u_1 \\ u_1 + u_2 \\ 0 \\ u_3 \end{bmatrix} = \begin{bmatrix} u_1 \\ u_2 \\ -u_1 - u_2 \\ u_3 \\ -u_3 \end{bmatrix} = \mathbf{u},$$

as predicted.

The level of a balance vector

A natural measure of the complexity of a balance vector is the number of its non-zero entries. For any $\mathbf{v}$ in $\text{Bal}_n(R)$ define the *level* of $\mathbf{v}$ to be the number of non-zero entries of $\mathbf{v}$, with the convention that the zero vector has level 1. Thus the level of a balance vector is a positive integer and the zero vector is the only balance vector with level 1. Clearly, if k is any integer satisfying $1 \leq k \leq n$, then $\mathrm{Bal}_n(R)$ has vectors of level k. One can think of the balance vectors as being classified in a hierarchy of levels. At level 1 is the zero vector, at level 2 the non-zero simple balance vectors, and thereafter balance vectors of increasing complexity.

It was shown in 2.3.1 that, provided $n > 1$, there is an R-basis of $\mathrm{Bal}_n(R)$ consisting of non-zero simple balance vectors, i.e., balance vectors of level 2. One can ask whether it is possible to find an R-basis of balance vectors of level k where k is any integer satisfying $1 < k \leq n$. The answer turns out to be affirmative.

(2.3.6). *Let R be an ordered domain and let k, n be integers satisfying $1 < k \leq n$. Then there is an R-basis of $\mathrm{Bal}_n(R)$ consisting of vectors of level k.*

Proof
Suppose that A is an $(n - 1) \times (n - 1)$ matrix with non-negative entries in R which has the following properties:

(a) $\det(A) = \pm 1$, where "det" denotes the determinant;

(b) each column of A has exactly $k - 1$ positive entries and the remaining $n - k$ entries are all 0.

The problem of finding such a matrix is postponed until later in the proof.

The next move is to adjoin an additional row to A, thereby creating an $n \times (n - 1)$ matrix A^*. Here the jth element of the nth row of A^* is defined to be the negative of the sum of the entries in column j of A. This implies that the sum of the entries in any column of A^* is 0, i.e., the columns of A^* are vectors in $\mathrm{Bal}_n(R)$. Notice also that each column of A^* has exactly k non-zero entries and hence has level k. We will show that the columns of A^* form an R-basis for $\mathrm{Bal}_n(R)$.

In order to establish this we let $\mathbf{y}$ be an arbitrary vector in R^{n-1}. Since $\det(A) \neq 0$, there is a unique vector $\mathbf{x}$ in R^{n-1} such that

$$A\mathbf{x} = \mathbf{y}.$$

Indeed $\mathbf{x} = A^{-1}\mathbf{y} = (\det(A))^{-1}\mathrm{adj}(A)\mathbf{y}$ where $\mathrm{adj}(A)$ is the adjoint of A, i.e. the transposed matrix of cofactors of A. Since entries of A belong to R and $\det(A) = \pm 1$, the vector $\mathbf{x}$ has all its entries in R.

Next form $\mathbf{y}^*$ from $\mathbf{y}$ in the same manner as A^* was formed from A; thus $\mathbf{y}^*$ is the n-column vector with entries $y_1, y_2, \ldots, y_{n-1}, -y_1 - \cdots - y_{n-1}$. We now claim that

$$A^*\mathbf{x} = \mathbf{y}^*.$$

To see this recall that $A\mathbf{x} = \mathbf{y}$ and note that the nth entry of $A^*\mathbf{x}$ equals

$$\sum_{j=1}^{n-1}\left(\sum_{i=1}^{n-1} -a_{ij}\right)x_j = -\sum_{i=1}^{n-1}\left(\sum_{j=1}^{n-1} a_{ij}x_j\right) = -\sum_{i=1}^{n-1} y_i,$$

which is the nth entry of $\mathbf{y}^*$. Now observe that $\mathbf{y}^*$ is actually a typical vector in $\mathrm{Bal}_n(R)$ since $y_1, \ldots, y_{n-1}$ are arbitrary. Also the equation $A^*\mathbf{x}^* = \mathbf{y}^*$ implies that $\mathbf{y}^*$ is a linear combination of the $n-1$ columns of A^*.

Finally, the columns of A^* are linearly independent because those of A are linearly independent — recall that $\det(A) \neq 0$. It follows that the $n-1$ columns of A^* constitute an R-basis of $\mathrm{Bal}_n(R)$; of course each column of A^* has level k.

There remains the problem of exhibiting an $(n-1)\times(n-1)$ matrix A satisfying (a) and (b): in fact there are many such matrices. One can start off with the $(k-1) \times (k-1)$ matrix

$$U = \begin{bmatrix} 1 & 2 & 1 & \cdots & 1 & 1 \\ 1 & 1 & 2 & \cdots & 1 & 1 \\ \vdots & \vdots & \vdots & \ddots & \vdots & \vdots \\ 1 & 1 & 1 & \cdots & 1 & 2 \\ 1 & 1 & 1 & \cdots & 1 & 1 \end{bmatrix}.$$

It is easy to compute its determinant by using row operations; in fact

$$\det(U) = (-1)^k.$$

The matrix U is used to construct a matrix A with the required properties by the following procedure.

First divide $n-1$ by $k-1$ to get a quotient q and a remainder r, both of which are integers; thus

$$n - 1 = (k-1)q + r$$

and $0 \le r < k-1$. The $(n-1) \times (n-1)$ matrix A is to have q blocks U down the main diagonal, with other entries 0 or 1 according to the following scheme:

$$
A = \begin{bmatrix}
\begin{array}{cccc|ccc}
U & & & & 1 & \cdots & 1 \\
 & U & & 0 & \vdots & & \vdots \\
0 & & \ddots & & 1 & \cdots & 1 \\
 & & & U & 0 & \cdots & 0 \\
 & & & & \vdots & \cdots & \vdots \\
 & & & & 0 & \cdots & 0 \\
\hline
 & & 0 & & & 1_r &
\end{array}
\end{bmatrix}
$$

Here the upper right hand block of 1's has size $(k-2) \times r$, while 1_r is the $r \times r$ identity matrix. Thus each column of A has exactly $k-1$ positive entries, with other entries being 0. Also $\det(A) = (\det(U))^q = (-1)^{kq}$. Thus A has all the required properties. $\qquad\square$

Example (2.3.2).

The previous result will now be illustrated by explicitly constructing a $\mathbb{Z}$-basis of $\mathrm{Bal}_6(\mathbb{Z})$ consisting of balance vectors of level 3. Thus $n = 6$ and $k = 3$ here. The matrix U is

$$
\begin{bmatrix} 1 & 2 \\ 1 & 1 \end{bmatrix}.
$$

Now $5 = 2 \times 2 + 1$, so $q = 2$ and $r = 1$. Assemble the matrix A as indicated above, to get

$$
A = \begin{bmatrix}
\begin{array}{cccc|c}
1 & 2 & 0 & 0 & 1 \\
1 & 1 & 0 & 0 & 0 \\
0 & 0 & 1 & 2 & 0 \\
0 & 0 & 1 & 1 & 0 \\
\hline
0 & 0 & 0 & 0 & 1
\end{array}
\end{bmatrix}.
$$

Notice that A has 2 positive entries in each column with other entries 0 and that $\det(A) = 1$, so (a) and (b) hold.

The final step is to add to A a sixth row whose entries are the negative column sums of A, which yields

$$A^* = \begin{bmatrix} 1 & 2 & 0 & 0 & 1 \\ 1 & 1 & 0 & 0 & 0 \\ 0 & 0 & 1 & 2 & 0 \\ 0 & 0 & 1 & 1 & 0 \\ 0 & 0 & 0 & 0 & 1 \\ -2 & -3 & -2 & -3 & -2 \end{bmatrix}.$$

The columns of A^* form a $\mathbb{Z}$-basis of $\mathrm{Bal}_6(\mathbb{Z})$ consisting of vectors of level 3.

Chapter Three

Transactions

Up to this point we have been concerned with developing the mathematical structures appropriate for describing the state of an accounting system at any instant, namely by balance vectors with entries belonging to an ordered domain. The next question to consider is how one can represent changes in the state of an accounting system which result from economic events. Such changes occur when a transaction is applied to the system, which means that there is a flow of value between accounts of the system. Some account balances will increase and others decrease. Of course, there may be some accounts which are unaffected by the transaction.

After a transaction has been applied to an accounting system, the system must still be in balance, i.e., the sum of all the account balances is zero. This implies that the sum of all the changes in account balances due to the transaction must equal zero. So the conclusion is that the effect of a transaction on an accounting system can be represented adding a fixed balance vector to the balance vector that describes the original state. Then the sum of these vectors is the balance vector representing the state of the system after the transaction has been applied.

This informal discussion indicates that balance vectors are also the key to understanding changes in the state of an accounting system, as well as the actual states of the system. The object of the present chapter is to give a formal treatment of transactions and their relation to balance vectors. Once established, this relationship permits the transfer of concepts and results for balance vectors to transactions.

3.1. Transaction Vectors

The formal definition of a transaction will now be given. Let n be a positive integer, which will correspond to the number of accounts in an accounting system, and let R be an ordered domain, which will be the realm of account values. Choose and fix a balance vector $\mathbf{v} \in \mathrm{Bal}_n(R)$. Then a function

$$\tau_{\mathbf{v}} : \mathrm{Bal}_n(R) \to \mathrm{Bal}_n(R)$$

is defined by the rule

$$\tau_{\mathbf{v}}(\mathbf{x}) = \mathbf{x} + \mathbf{v}, \ (\mathbf{x} \in \mathrm{Bal}_n(R)).$$

Thus the function $\tau_{\mathbf{v}}$ simply adds the fixed balance vector $\mathbf{v}$ to each balance vector $\mathbf{x}$; of course $\mathbf{x} + \mathbf{v}$ is also a balance vector. The function $\tau_{\mathbf{v}}$ is called the *transaction* corresponding to the balance vector $\mathbf{v}$. The set of all such transactions will be denoted by

$$\mathrm{Trans}_n(R) = \{\tau_{\mathbf{v}} \mid \mathbf{v} \in \mathrm{Bal}_n(R)\}.$$

Notice that the zero vector $\mathbf{0}$ corresponds to the identity function since $\tau_{\mathbf{0}}(\mathbf{x}) = \mathbf{x} + \mathbf{0} = \mathbf{x}$. Thus $\tau_{\mathbf{0}}$ is the *identity transaction*, which causes no change in the system.

Recall that two functions α, β from a set to itself can be combined by using functional composition to yield a new function, the *composite*

$$\alpha \circ \beta,$$

defined by $\alpha \circ \beta(x) = \alpha(\beta(x))$. In the case of transactions $\tau_{\mathbf{v}}$ and $\tau_{\mathbf{w}}$, observe that $\tau_{\mathbf{v}} \circ \tau_{\mathbf{w}}$ sends $\mathbf{x} \in \mathrm{Bal}_n(R)$ to $(\mathbf{x} + \mathbf{w}) + \mathbf{v} = \mathbf{x} + (\mathbf{v} + \mathbf{w})$, as does $\tau_{\mathbf{w}} \circ \tau_{\mathbf{v}}$. Therefore

$$\tau_{\mathbf{v}} \circ \tau_{\mathbf{w}} = \tau_{\mathbf{v}+\mathbf{w}} = \tau_{\mathbf{w}} \circ \tau_{\mathbf{v}}.$$

Hence $\tau_{\mathbf{v}} \circ \tau_{-\mathbf{v}} = \tau_{\mathbf{0}}$ and $\tau_{-\mathbf{v}}$ is the inverse of the transaction $\tau_{\mathbf{v}}$.

The above equations, together with the associative law of functional composition, show that $\mathrm{Trans}_n(R)$ is an abelian group. There is also an R-module structure on $\mathrm{Trans}_n(R)$: for one can define $r\tau_{\mathbf{v}}$, where $r \in R$ and $\mathbf{v} \in \mathrm{Bal}_n(R)$, by the rule

$$r\tau_{\mathbf{v}} = \tau_{r\mathbf{v}}.$$

Thus $(r\tau_{\mathbf{v}})(\mathbf{x}) = \mathbf{x} + r\mathbf{v}$. It is very easy to check the module axioms in 2.2, so that $\mathrm{Trans}_n(R)$ is an R-module.

By this point it should be apparent that the modules $\mathrm{Trans}_n(R)$ and $\mathrm{Bal}_n(R)$ are very similar. This may be formalized by saying that these R-modules are *isomorphic*, meaning that there is an *isomorphism*, or bijective homomorphism, between them.

(3.1.1). *The assignment* $\mathbf{v} \mapsto \tau_{\mathbf{v}}$ *determines a function*

$$\tau : \mathrm{Bal}_n(R) \to \mathrm{Trans}_n(R)$$

which is an isomorphism of R-modules.

Proof
It is perfectly clear that τ is a bijection. Equations $\tau_{\mathbf{v}+\mathbf{w}} = \tau_{\mathbf{v}} \circ \tau_{\mathbf{w}}$ and $r\tau_{\mathbf{v}} = \tau_{r\mathbf{v}}$ show that $\tau(\mathbf{v}+\mathbf{w}) = \tau(\mathbf{v}) \circ \tau(\mathbf{w})$ and $\tau(r\mathbf{v}) = r\tau(\mathbf{v})$, so that τ is a homomorphism of R-modules. $\square$

In the interests of brevity we will often not distinguish between the transaction $\tau_{\mathbf{v}}$ and the corresponding balance vector $\mathbf{v}$, referring to either as a *transaction vector*. Thus we can speak of *simple transactions* and *elementary transactions*, corresponding to simple balance vectors and elementary balance vectors. Note that a simple transaction is one with level 2 and is just an exchange of value between two accounts, all other accounts being unaffected.

The essential feature of an isomorphism between two modules is that it preserves all structural properties of the modules. Thus all the properties of $\mathrm{Bal}_n(R)$ which were established in Chapter 2 may be transferred to $\mathrm{Trans}_n(R)$. For example, one can conclude on the basis of 2.3.1 that $\mathrm{Trans}_n(R)$ is a free R-module of rank $n-1$. One can also speak of the *level* of a transaction $\tau_{\mathbf{v}}$, meaning thereby the number of non-zero entries in the balance vector $\mathbf{v}$.

It is worthwhile stating explicitly the analog of 2.3.6 for $\mathrm{Trans}_n(R)$ since it furnishes information about transactions.

(3.1.2). *Let* n, k *be integers satisfying* $1 < k \leq n$. *Then* $\mathrm{Trans}_n(R)$ *has an R-basis consisting of transactions of level* k.

Applying 3.1.2 and the equation $r\tau_{\mathbf{v}} = \tau_{r\mathbf{v}}$, we deduce:

(3.1.3). *Every transaction on a set of* n *accounts is the composite of a sequence of transactions of level* k *where* k *is any integer such that* $1 < k \leq n$.

Specializing 3.1.3 to the case $k = 2$, we obtain:

(3.1.4). *Every transaction is a composite of simple transactions.*

This means that any transaction, however complex, can be effected by a suitable sequence of exchanges between pairs of accounts.

Transactions and T-diagrams

In accounting a common way of representing the transactions that have been applied to an accounting system is by means of what are called *T-diagrams*. Each account has a T-diagram which lists the debits and credits that have been applied to the account in two columns, the column on the left giving the debits and the one on the right the credits. The T-diagram for account a_i over some accounting period has the general form

Account a_i

x_{i1}	y_{i1}
x_{i2}	y_{i2}
$\vdots$	$\vdots$
x_{ik_i}	y_{im_i}

where all x_{ij}, y_{ij} are positive.

By convention in accounting a transaction which increases the value of an asset account such as a bank account is said to *debit* the account; if it decreases the value of such an account, then it *credits* the account. Similarly, if a transaction increases the balance of a liability account, then it *credits* the account and if it decreases the balance of such an account, it *debits* the account. This reversal of the common usage of the terms "debit" and "credit" reflects the fact that amounts in an asset account such as cash are owed by the company to the owners, whereas an amount in a liability account such as a mortgage is owed to the company by the owners. For example, a transaction that is a sale might debit the bank account and credit inventory. Repayment of a loan will credit the bank account and debit bank loan.

At this point the advantage of the balance vector representation of transactions and balances becomes apparent since the signs of the entries in the balance vector take care of both asset and liability accounts. A transaction debits an account if, *allowing for the signs of the vector entries*, it increases the corresponding entry in the balance vector, and it credits the account if it decreases the entry in the balance vector.

Remembering that the left hand column of a T-diagram represents debits and the right hand one credits, we conclude from the T-diagram above that the balance of the the ith account will increase over the period by an amount

$$x_{i1} + x_{i2} + \cdots + x_{ik_i} - y_{i1} - y_{i2} - \cdots - y_{im_i}.$$

(Of course, if this amount is negative, it represents a decrease in the account balance).

It is possible to enrich the T-diagrams by adding a time component. Then x_{ij} is replaced by a pair (t_{ij}, x_{ij}) where t_{ij} is the expiration time for the jth transaction to be applied to the ith account. This procedure will be elaborated on in the discussion of automata in Chapter 6.

T-diagrams and the Pacioli group

Consider once again the T-diagram of the ith account of an accounting system. Let d and c denote the respective sums of the debit and credit columns in the diagram. Notice that $d - c$, the increase in value of the ith account over the period, is not affected if we replace d by $d+x$ and c by $c+x$: it is only the difference between d and c that matters. Motivated by this simple observation, let us consider a relation E on the set product $\mathbb{Z} \times \mathbb{Z}$ of all ordered pairs (a, b), $(a, b \in \mathbb{Z})$, defined by

$$(a, b)\ E\ (a', b') \quad \text{if and only if} \quad a - b = a' - b';$$

it is easy to check that E is an equivalence relation on $\mathbb{Z} \times \mathbb{Z}$. Write

$$[a, b]$$

for the E-equivalence class containing (a, b) and let P be the set of all such equivalence classes. The next step is to define a binary operation on P:

$$[a, b] + [a', b'] = [a + a', b + b'].$$

Again it is a simple matter to see that this operation is well-defined, i.e., it depends on the classes $[a, b]$, $[a', b']$, not on the representing ordered pairs. Next we quickly verify that Properties 1-4 in 2.1 hold, which means that P is an abelian group with respect to the operation just defined; note that the identity element is $[0, 0]$, which is the same as $[a, a]$ for any $a \in \mathbb{Z}$. Following Ellerman [1985], we shall call P the *Pacioli group* because of its connection with the double entry accounting system pioneered by Pacioli. The identity of this group is not a mystery: for the assignment

$$[a, b] \mapsto a - b, \ (a, b \in \mathbb{Z}),$$

from P to the additive group of integers $\mathbb{Z}$ is quickly seen to yield an isomorphism $\alpha : P \to \mathbb{Z}$: for it is clearly a bijection and

$$\alpha([a, b] + [a', b']) = \alpha([a, b]) + \alpha([a', b']).$$

Thus in effect the groups P and $\mathbb{Z}$ have identical properties.

Returning to the T-diagram of the ith account, we note that it determines the element $[d, c]$ of the Pacioli group. Thus at any instant each account, through its T-diagram, determines an element of this group. However, in passing from T-diagrams to the elements of the Pacioli group it is apparent that much information about the transactions has been lost, a fact that obviously limits the usefulness of the construction.

The balance matrix

The history of an accounting system over some period of time can be described by listing the T-diagrams for the accounts. An alternative way to describe this history is by listing in order the successive balance vectors of the system after each transaction has been applied. These balance vectors can be used as the columns of a matrix M. Since the matrix M determines the balance sheet of the company, we call it the *balance matrix* of the accounting system.

Notice that we can recover the transactions which have been applied to the system from the balance matrix M by subtracting successive columns. In more detail, let $\mathbf{b}(0)$ be the initial balance vector of the system and let the successive balance vectors after application of k transactions $\mathbf{v}(1), \mathbf{v}(2), \ldots, \mathbf{v}(k)$ be $\mathbf{b}(1), \mathbf{b}(2), \ldots, \mathbf{b}(k)$; thus

$$\mathbf{b}(i) = \mathbf{b}(i - 1) + \mathbf{v}(i)$$

and the balance matrix over the period is

$$M = [\mathbf{b}(0), \mathbf{b}(1), \ldots, \mathbf{b}(k)].$$

Then we recover the transactions from the matrix M from the equations

$$\mathbf{v}(i) = \mathbf{b}(i) - \mathbf{b}(i-1).$$

Finally, given the balance matrix M we can also reconstruct all the T-diagrams. The procedure is first to find the successive transactions that have been applied, as already explained. To obtain the T-diagram of the ith account, identify the positive i-entries in the transaction vectors and record them in the debit column of the T-diagram; then put the negative i-entries in the credit column. Thus the balance matrix determines the set of T-diagrams uniquely. We illustrate the procedure with an example.

Example (3.1.1).

Suppose that an accounting system with four accounts has balance matrix over some period

$$\begin{bmatrix} 500 & 400 & 300 & 700 \\ 100 & 100 & -50 & 50 \\ -350 & -300 & -100 & -400 \\ -250 & -200 & -150 & -350 \end{bmatrix}.$$

To construct the four T-diagrams first find the three transactions that have been applied:

$$\begin{bmatrix} -100 \\ 0 \\ 50 \\ 50 \end{bmatrix}, \quad \begin{bmatrix} -100 \\ -150 \\ 200 \\ 50 \end{bmatrix}, \quad \begin{bmatrix} 400 \\ 100 \\ -300 \\ -200 \end{bmatrix}.$$

Now read off the T-diagram of accounts a_1 and a_2 as

Account a_1

400	100
	100

and

Account a_2

100	150

Similarly for the other accounts.

3.2. Transaction Types

A problem which will be discussed in subsequent chapters is to determine the appropriateness of applying a given transaction to an accounting system. In deciding this question it is sometimes the arrangement of credits and debits, i.e., the signs of the entries in the transaction vector, that are important. The actual entries may be of lesser significance. This is the background to the definition of type.

Let n be a positive integer and R an ordered domain. The *type* of a balance vector $\mathbf{v} \in \mathrm{Bal}_n(R)$,

$$\mathrm{type}(\mathbf{v}),$$

is the n-column vector whose ith entry is 0, $+$ or $-$ according as $v_i = 0$, $v_i > 0$ or $v_i < 0$ respectively. For example, if $n = 4$ and

$$\mathbf{v} = \begin{bmatrix} -300 \\ 400 \\ -100 \\ 0 \end{bmatrix} \in \mathrm{Bal}_4(\mathbb{Z}),$$

the type of $\mathbf{v}$ is

$$\mathrm{type}(\mathbf{v}) = \begin{bmatrix} - \\ + \\ - \\ 0 \end{bmatrix}.$$

The *type of a transaction* $\tau_{\mathbf{v}}$ is then defined to be the type of $\mathbf{v}$.

The identity transaction has type $\mathbf{0}$, while the type of a simple transaction contains a single $+$ and $-$, with other entries 0. Notice that any non-zero transaction type must have at least one $+$ and at least one $-$.

A partial order on transaction types

Let $\mathbf{s}$ and $\mathbf{t}$ be types of balance vectors in $\mathrm{Bal}_n(R)$: thus the entries of the vectors $\mathbf{s}, \mathbf{t}$ are 0, $+$ or $-$. A binary relation $\leq$ on the set of types of vectors in $\mathrm{Bal}_n(R)$ is defined as follows:

$$\mathbf{s} \leq \mathbf{t}$$

is to mean that $s_i = t_i$ or $s_i = 0$ for $i = 1, 2, \ldots, n$. Thus $\mathbf{s}$ and $\mathbf{t}$ have the same configuration of $+$ and $-$ signs except that $\mathbf{s}$ may have more zeros. It is a simple matter to verify that this relation is reflexive, transitive and antisymmetric, so it is a *partial order* on the set of types. But this is not a linear order since not every pair of types is comparable: for example, the types

$$\begin{bmatrix} - \\ + \\ - \\ 0 \end{bmatrix} \quad \text{and} \quad \begin{bmatrix} - \\ - \\ + \\ 0 \end{bmatrix}$$

are incomparable.

As is usually done with a partial order, one can visualize the partially ordered set of types by means of its *Hasse diagram*, in which the least complex types occur lower down in the diagram. At the lowest point will be the type of the zero vector $\mathbf{0}$, which consists entirely of zeros, while $\text{type}(\mathbf{v})$ sits directly below $\text{type}(\mathbf{w})$ if the entries of $\text{type}(\mathbf{v})$ and $\text{type}(\mathbf{w})$ are the same except that $\text{type}(\mathbf{v})$ has one more zero entry.

A related concept is that of level. The *level of a transaction* is defined to be the level of its associated balance vector as defined in 2.3, i.e., it is the number of $+$ and $-$ signs in the type. The concept of level permits a *linear* ordering of transaction types in which types of small level occur further down in the partial ordering of types. Clearly the highest possible level is n and the lowest level is 1, the type of the identity transaction. Thus the level is to be regarded as a measure of the complexity of a transaction type.

Example (3.2.1).

There are 13 transaction types in $\text{Trans}_3(R)$. These are listed below in descending order of levels:

$$\text{level 3}: \quad \begin{bmatrix} + \\ + \\ - \end{bmatrix}, \begin{bmatrix} - \\ + \\ + \end{bmatrix}, \begin{bmatrix} + \\ - \\ + \end{bmatrix}, \begin{bmatrix} - \\ - \\ + \end{bmatrix}, \begin{bmatrix} + \\ - \\ - \end{bmatrix}, \begin{bmatrix} - \\ + \\ - \end{bmatrix}$$

$$\text{level 2}: \quad \begin{bmatrix} + \\ - \\ 0 \end{bmatrix}, \begin{bmatrix} 0 \\ + \\ - \end{bmatrix}, \begin{bmatrix} - \\ 0 \\ + \end{bmatrix}, \begin{bmatrix} - \\ + \\ 0 \end{bmatrix}, \begin{bmatrix} 0 \\ - \\ + \end{bmatrix}, \begin{bmatrix} + \\ 0 \\ - \end{bmatrix}$$

$$\text{level 1}: \quad \begin{bmatrix} 0 \\ 0 \\ 0 \end{bmatrix}$$

As an illustration of the partial ordering of types, observe that

$$\begin{bmatrix} + \\ 0 \\ - \end{bmatrix} \leq \begin{bmatrix} + \\ + \\ - \end{bmatrix}.$$

In the daily operation of a real-life accounting system many of the transactions applied will likely be at a low level. For example, funds might be moved between two accounts, which is a transaction of level 2. In the case of a retail firm, a sale might involve debiting cash, crediting inventory and crediting profit and loss, a transaction of level 3. Nevertheless one can envisage transactions which occur at a high level and so are of complex type. For example, funds might be disbursed from cash to several employee payroll or pension accounts. Thus transaction types of high level are a distinct possibility.

Combinatorial properties of transaction types

An examination of the transaction types in $\text{Trans}_3(R)$ in Example 3.1.2 above raises some combinatorial questions about types. For example, it is natural to ask how many transaction types there are in $\text{Bal}_n(R)$, and one might also enquire about the number of types of given level. Finally, there is the more ambitious question: which level contains the largest number of transaction types? We shall derive some simple formulas which answer these questions. Aside from their intrinsic interest, these combinatorial questions provide insight into the relative complexity of transaction types at different levels. In this result $\binom{n}{r}$ denotes the binomial coefficient

$$\frac{n(n-1)\cdots(n-r+1)}{r!}.$$

(3.2.1). *Let n be an integer greater than 1.*

1. *The number of n-transaction types of level r is $\binom{n}{r}(2^r - 2)$, where $1 < r \leq n$.*

2. *The total number of n-transaction types is $3^n - 2^{n+1} + 2$.*

Proof

1. In order to construct an n-transaction type of level r one must first pick the r "slots" in which a $+$ or $-$ is to be placed; this may be done in $\binom{n}{r}$ ways. Then one has to count the number of ways of placing a $+$ or $-$ in each of the r chosen slots, taking care not to have a $+$ in every slot or a $-$ in every slot. This can be done in $2^r - 2$ ways. The remaining $n - r$ slots get 0's, so the type has been determined. Hence the number of types at level r is $\binom{n}{r}(2^r - 2)$.

2. The total number of n-transaction types is the sum of the numbers of types at levels 1 through n. Since there is just one type of level 1, this is

$$1 + \sum_{r=2}^{n} \binom{n}{r}(2^r - 2) = 1 + \sum_{r=2}^{n} \binom{n}{r} 2^r - 2 \sum_{r=2}^{n} \binom{n}{r}.$$

Now by the Binomial Theorem

$$\sum_{r=0}^{n} \binom{n}{r} 2^r = (1 + 2)^n = 3^n$$

and

$$\sum_{r=0}^{n} \binom{n}{r} = (1 + 1)^n = 2^n.$$

Hence the total number of types is

$$1 + (3^n - 1 - 2n) - 2(2^n - 1 - n) = 3^n - 2^{n+1} + 2. \qquad \square$$

A similar combinatorial problem arises when one asks for the number of transaction types with a specified number of debits or credits.

(3.2.2). *Let n and r be integers such that $1 \le r \le n$. Then*

1. *the number of n-transaction types with exactly r entries $+$, i.e., debits, is $\binom{n}{r}(2^{n-r} - 1)$, and this is also the number of types with exactly r entries $-$, i.e., credits;*

2. *the number of n-transaction types with exactly r zeros (i.e., r unaffected accounts) is $\binom{n}{r}(2^{n-r} - 2)$ if $r < n$ and 1 if $r = n$.*

Proof
1. Choose the r slots which are to receive a $+$ sign in $\binom{n}{r}$ ways. Then place a 0 or a $-$ in the remaining $n - r$ slots, but do not put a 0 in every slot: for there must be at least one $-$ in the type. This can be done in $2^{n-r} - 1$ ways. So the number of types with exactly r $+$ signs is $\binom{n}{r}(2^{n-r} - 1)$. The same argument handles the case of r minus signs.

2. Let $r < n$ and choose the r slots to receive 0's in $\binom{n}{r}$ ways. Then place a $+$ or $-$ in each of the remaining $n - r$ slots, with at least one $+$ sign and one $-$ sign. This can be done in $2^{n-r} - 2$ ways, so the answer is $\binom{n}{r}(2^{n-r} - 2)$. If $r = n$, the answer is clearly 1.   □

The level with the largest number of types

Let n be a fixed positive integer. Another natural question is: which level has the largest number of n-transaction types? Clearly one can assume that $n > 1$ here. Then by 3.2.1 the problem is to find the integer r satisfying $1 \le r \le n$ for which the integer $\binom{n}{r}(2^r - 2)$ achieves its maximum value. Now it is well-known that the maximum value of $\binom{n}{r}$ occurs at $r = \left[\frac{n}{2}\right]$ (the largest integer $\le \frac{n}{2}$). However, one would expect the maximum value of $\binom{n}{r}(2^r - 2)$ to occur for a larger value of r because of the presence of the factor $2^r - 2$. In fact the answer is roughly $\frac{2}{3}n$, as the next result indicates:

(3.2.3). *Let n be a positive integer. Then the largest number of n-transaction types occurs at level*

$$\left[\frac{2n + 2}{3}\right].$$

For example, when $n = 100$, the largest number of types occurs at level 67, and the number of types at that level is $\binom{100}{67}(2^{67} - 2)$.

Proof

It may be assumed that $n > 1$. By 3.2.1 we need to determine the value of r which makes the integer $a_r = \binom{n}{r}(2^r - 2)$ largest; here n is fixed and $1 < r < n$. Now

$$a_{r+1} - a_r = \frac{n!}{(r + 1)!\,(n - r)!}\left(n(2^{r+1} - 2) - (3r + 1)2^r + 4r + 2\right).$$

Therefore $a_r < a_{r+1}$ if and only if n exceeds the number

$$b_r = \frac{(3r+1)2^r - 4r - 2}{2^{r+1} - 2}.$$

It is easy to see that $\{b_r\}$ is a strictly increasing sequence of positive numbers. Also, $\lim_{r\to\infty} b_r = +\infty$, so there is a least r such that $n \leq b_r$ and this r will be a level with the largest number of transaction types. Thus we have to find an integer r such that

$$b_{r-1} < n \leq b_r.$$

Put

$$\varepsilon_i = \frac{3i+1}{2} - b_i;$$

then a short calculation shows that

$$\varepsilon_i = \frac{i+1}{2^{i+1} - 2}$$

for $i \geq 1$. Of course $\lim_{i\to\infty} \varepsilon_i = 0$ and in fact $0 < \varepsilon_i \leq \frac{1}{2}$ if $i \geq 2$. On writing $b_i = \frac{3i+1}{2} - \varepsilon_i$, the inequality $b_{r-1} < n \leq b_r$ becomes

$$\frac{3r-2}{2} - \varepsilon_{r-1} < n \leq \frac{3r+1}{2} - \varepsilon_r.$$

Solving for r, one finds that

$$\frac{2n-1}{3} + \frac{2}{3}\varepsilon_r \leq r < \frac{2n+2}{3} + \frac{2}{3}\varepsilon_{r-1}. \qquad (*)$$

There are now three cases to consider. Suppose first that $n \equiv 0$ (mod 3) and $n = 3k$. Then $(*)$ yields

$$2k - \frac{1}{3} + \frac{2}{3}\varepsilon_r \leq r < 2k + \frac{2}{3} + \frac{2}{3}\varepsilon_{r-1}.$$

If $r = 2$, then $n = 3$ and the result is true by Example 3.2.1, while $r = 1$ is impossible; thus we may assume that $r \geq 3$. Since $\varepsilon_{r-1} \leq \frac{1}{2}$, the last inequality becomes $2k \leq r < 2k + 1$, so that $r = 2k = [\frac{2n+2}{3}]$.

Next suppose that $n \equiv 1$ (mod 3) and write $n = 1 + 3k$. Then $(*)$ yields

$$2k + \frac{1}{3} + \frac{2}{3}\varepsilon_r \leq r < 2k + \frac{4}{3} + \frac{2}{3}\varepsilon_{r-1}.$$

Since $\varepsilon_{r-1} \leq \frac{1}{2}$, this gives $r = 2k + 1 = \lceil \frac{2n+2}{3} \rceil$.

Finally, if $n \equiv 2 \pmod 3$ and $n = 2 + 3k$, then we have by $(*)$

$$2k + 1 + \frac{2}{3}\varepsilon_r \leq r < (2k + 2) + \frac{2}{3}\varepsilon_{r-1},$$

which yields $r = 2k + 2 = \lceil \frac{2n+2}{3} \rceil$ since one can assume $n > 2$. $\square$

Observe that 3.2.3 leaves open the possibility that the maximum number of types occurs at two successive levels, and indeed this happens when $n = 3$, as Example 3.2.1 above shows. However this is the only time it happens. The reason is that $a_r = a_{r+1}$ holds only if $n = b_r$ and it is easy to see that b_r is an integer only when $r = 1$ or 2, which is consistent only with $n = 3$.

3.3. Transactions, Matrices and Digraphs

Up to this point our favored method of representing transactions has been by balance vectors over an ordered domain. However some years ago Mattessich, in a well-known paper [Mattesich 1957], gave an ingenious method of representing transactions on an accounting system by square matrices with non-negative integral entries. It is instructive to compare Mattessich's matrix method with the current approach. A comparison of the two methods will highlight some of the main features of the balance vector technique and its advantages. A detailed analysis of the relation between the methods is presented in this section.

From matrices to transactions

Let n be a positive integer and R an ordered domain, and let $M = [m_{ij}]$ be an $n \times n$ matrix over R, so that the entries of M lie in R. An n-column vector $\mathbf{v}$ is formed from M by the following procedure: the ith entry of the vector $\mathbf{v}$ is given by

$$v_i = \sum_{j=1}^{n}(m_{ij} - m_{ji}).$$

Thus the value of the ith account a_i is debited, i.e., increased by amount m_{ij} and credited, i.e., decreased by m_{ji}. Otherwise stated, the ith entry of $\mathbf{v}$ is obtained from M by forming the sum of the entries in row i and subtracting from it the sum of the entries in

column i. The formula is valid even if some of the matrix entries
are negative.

. The critical observation is that $\mathbf{v}$ is a balance vector, the reason
being that in the sum $\sum_{i=1}^{n} v_i$ each m_{ij} occurs twice, once with a
positive sign and once with a negative sign. Thus the matrix M
determines a transaction $\tau_{\mathbf{v}}$. Finally, notice that diagonal entries
m_{ii} have no effect on the vector $\mathbf{v}$ since they cancel in the sum
expressing v_i.

Example (3.3.1).

Consider the 4×4 matrix over $\mathbb{Z}$

$$M = \begin{bmatrix} 0 & 100 & -200 & 750 \\ 400 & -50 & 100 & 0 \\ 100 & 0 & 50 & 100 \\ -400 & -100 & 100 & 0 \end{bmatrix}.$$

Following the row-sum minus column-sum rule, we find that the
corresponding transaction is represented by the vector

$$\mathbf{v} = \begin{bmatrix} 550 \\ 500 \\ 200 \\ -1250 \end{bmatrix}.$$

Example (3.3.2).

Let
$$E(i, j)$$
denote the $n \times n$ *elementary matrix* whose (i, j) entry is 1 and whose
other entries are all 0; here $i \neq j$. It is an important observation that
*the elementary matrix $E(i, j)$ determines the elementary transaction
vector* $\mathbf{e}(i, j)$, which has ith entry $+1$, jth entry -1 and other entries
zero: in particular this is a simple transaction.

The Mattessich function

The procedure just described for associating a balance vector
with a matrix can be formalized by means of a function μ. Let

$$M_n(R)$$

be the set of all $n \times n$ matrices over an ordered domain R. Matrix algebra provides natural operations of addition and scalar multiplication by elements of R for the set $M_n(R)$. Furthermore the laws of matrix algebra guarantee that $M_n(R)$, like $\mathrm{Bal}_n(R)$, is an R-module.

Mattessich's procedure yields a function

$$\mu : M_n(R) \to \mathrm{Bal}_n(R)$$

which is given by the rule that follows: if $M = [m_{ij}] \in M_n(R)$, then $\mu(M)$ is the n-column vector whose ith entry is

$$\sum_{j=1}^{n}(m_{ij} - m_{ji}).$$

The function μ will be called the *Mattessich function*.[1] Keep in mind that there is a module isomorphism from $\mathrm{Bal}_n(R)$ to $\mathrm{Trans}_n(R)$, so that, on composing μ with this, we obtain another module isomorphism

$$\mu' : M_n(R) \to \mathrm{Trans}_n(R).$$

There is a simple description of the function μ in terms of matrix products. Denote by I the n-column vector with all its entries equal to 1,

$$I = \begin{bmatrix} 1 \\ 1 \\ \vdots \\ 1 \end{bmatrix}.$$

Then by direct matrix multiplication we see that the ith-entry of the column vector $MI - M^T I$ is exactly $\sum_{j=1}^{n}(m_{ij} - m_{ji})$, where M^T is the transpose of the matrix M. The point to note here is that right multiplication of a matrix by I sums the elements in each row of the matrix.

The result of this observation is a simple formula for the Mattessich function:

$$\mu(M) = (M - M^T)I.$$

Use of this formula and elementary matrix algebra show that

$$\mu(M + N) = \mu(M) + \mu(N) \quad \text{and} \quad \mu(rM) = r\mu(M),$$

---

[1]Actually Mattessich used a slightly different procedure, with the roles of m_{ij} and m_{ji} reversed in the definition of μ.

where $M, N \in M_n(R)$ and $r \in R$. Thus μ is a homomorphism of R-modules.

The basic properties of the Mattessich function μ, and by implication of the function μ', are summarized in the following result:

(3.3.1). *Let n be a positive integer and R an ordered domain. Then*

1. *$\mu(M) = (M - M^T)I$ where I is the n-column vector with all entries equal to 1.*

2. *$\mu : M_n(R) \rightarrow \mathrm{Bal}_n(R)$ is a surjective homomorphism of R-modules.*

Proof
Only the surjectivity of μ requires a comment: clearly we can assume that $n > 1$. Recall that every balance vector $\mathbf{v}$ can be written in the form $\mathbf{v} = r_1\mathbf{e}(1,2) + \cdots + r_{n-1}\mathbf{e}(n-1,n)$ where $r_i \in R$. Also $\mathbf{e}(i, i+1) = \mu(E(i, i+1))$, as was pointed out in Example 3.3.2 above. Therefore

$$\mu(r_1 E(1,2) + \cdots + r_{n-1}E(n-1,n)) = \mathbf{v}$$

and μ is surjective. $\square$

Of course this result shows that every balance vector arises from some matrix, so Mattessich's representation of transactions is effective every transaction.

While the matrix representation of transactions is very elegant, it does have some disadvantages. Firstly, it is somewhat unwieldy: an $n \times n$ matrix has n^2 entries whereas a balance vector is determined by only $n-1$ parameters. Then, because of this built-in redundancy, different matrices can determine the same balance vector, and hence the same transaction.

These defects can be remedied by restricting attention to matrices all of whose non-zero entries lie on the superdiagonal, (where $n > 1$),

$$M = \begin{bmatrix} 0 & m_1 & 0 & \cdots & 0 & 0 \\ 0 & 0 & m_2 & \cdots & 0 & 0 \\ 0 & 0 & 0 & \cdots & 0 & 0 \\ \vdots & \vdots & \vdots & \ddots & \vdots & \vdots \\ 0 & 0 & 0 & \cdots & 0 & m_{n-1} \\ 0 & 0 & 0 & \cdots & 0 & 0 \end{bmatrix}$$

Notice that for this superdiagonal matrix

$$\mu(M) = \begin{bmatrix} m_1 \\ m_2 - m_1 \\ \vdots \\ m_{n-1} - m_{n-2} \\ -m_{n-1} \end{bmatrix},$$

which equals $m_1 \mathbf{e}(1,2) + m_2 \mathbf{e}(2,3) + \cdots + m_{n-1} \mathbf{e}(n-1, n)$. Since this is a general vector of $\text{Bal}_n(R)$, every balance vector, and hence every transaction, actually arises from a superdiagonal matrix.

This observation provides the motivation for introducing a function

$$\lambda : \text{Bal}_n(R) \to M_n(R),$$

which is defined by the rule that

$$\lambda(m_1 \mathbf{e}(1,2) + m_2 \mathbf{e}(2,3) + \cdots + m_{n-1} \mathbf{e}(n-1, n))$$

is to equal

$$M = \begin{bmatrix} 0 & m_1 & 0 & \cdots & 0 & 0 \\ 0 & 0 & m_2 & \cdots & 0 & 0 \\ 0 & 0 & 0 & \cdots & 0 & 0 \\ \vdots & \vdots & \vdots & \ddots & \vdots & \vdots \\ 0 & 0 & 0 & \cdots & 0 & m_{n-1} \\ 0 & 0 & 0 & \cdots & 0 & 0 \end{bmatrix}$$

In particular notice that $\lambda(\mathbf{e}(i, i+1)) = E(i, i+1)$.

It follows at once from the definition that λ is a homomorphism of R-modules. Also $\mu \circ \lambda$ is the identity function on $\text{Bal}_n(R)$, so that λ is a right inverse of μ and λ is therefore injective. Thus for any $\mathbf{v}$ in $\text{Bal}_n(R)$, we have $\mathbf{v} = \mu \circ \lambda(\mathbf{v}) = \mu(A)$ where $A = \lambda(\mathbf{v})$. Consequently every transaction $\tau_{\mathbf{v}}$ arises from a superdiagonal matrix obtained by applying the function μ' to $\lambda(\mathbf{v})$.

In fact each transaction arises from a unique superdiagonal matrix. For if $\mathbf{v} = \mu(M_1) = \mu(M_2)$, where M_1 and M_2 are superdiagonal, then $\mu(M_1 - M_2) = 0$. Since $M_1 - M_2$ is also superdiagonal, it follows from the equation for $\mu(M)$ that $M_1 = M_2$. This is stated formally as:

(3.3.2). *Every balance vector $\mathbf{v}$ in $\text{Bal}_n(R)$ is uniquely expressible in the form $\mu(M)$ where M is a superdiagonal $n \times n$ matrix; moreover $M = \lambda(\mathbf{v})$.*

Example (3.3.3).

Consider the balance vector

$$\mathbf{v} = \begin{bmatrix} 100 \\ -300 \\ -300 \\ 500 \end{bmatrix}.$$

First we express $\mathbf{v}$ in terms of elementary balance vectors as in 2.3.1:

$$\mathbf{v} = 100\mathbf{e}(1,2) - 200\mathbf{e}(2,3) - 500\mathbf{e}(3,4).$$

The superdiagonal matrix corresponding to $\mathbf{v}$ is therefore

$$\lambda(\mathbf{v}) = \begin{bmatrix} 0 & 100 & 0 & 0 \\ 0 & 0 & -200 & 0 \\ 0 & 0 & 0 & -500 \\ 0 & 0 & 0 & 0 \end{bmatrix}.$$

Matrices with non-negative entries

In his original article Mattessich dealt only with matrices having non-negative entries. As Example 3.3.3 shows, the superdiagonal matrix of a transaction can have negative entries. This is easily corrected by switching negative entries to the subdiagonal and changing the signs of such entries.

Let $\mathbf{v} \in \mathrm{Bal}_n(R)$; if the $(i, i+1)$ entry of the superdiagonal matrix $\lambda(\mathbf{v})$ is negative, say $-d$ with $d > 0$, replace it by 0 and put a $+d$ in the $(i+1, i)$ position. This procedure does not change difference between the row-sum and the column-sum, so that the resulting matrix, say

$$\lambda^*(\mathbf{v}),$$

is still mapped to $\mathbf{v}$ by μ; also it has all its non-zero entries positive and they lie on the superdiagonal or subdiagonal.

Thus in the previous example

$$\lambda^*(\mathbf{v}) = \begin{bmatrix} 0 & 100 & 0 & 0 \\ 0 & 0 & 0 & 0 \\ 0 & 200 & 0 & 0 \\ 0 & 0 & 500 & 0 \end{bmatrix}.$$

However the function λ^*, unlike λ and μ, is not a homomorphism.

The relationship between the functions λ and μ is clarified in the next result.

(3.3.3). *The functions λ and μ have the following properties.*

1. $\mathrm{Bal}_n(R) = \mathrm{Im}(\lambda) \oplus \mathrm{Ker}(\mu)$;

2. $\mathrm{Im}(\lambda)$ *consists of all $n \times n$ superdiagonal matrices over R;*

3. A matrix M belongs to $\mathrm{Ker}(\mu)$ if and only if $2M = S_1 + S_2$ where S_1 is symmetric and S_2 is skew symmetric with all its row sums equal to 0.

Proof

1. Recall that $\mu \circ \lambda$ is the identity function on $\mathrm{Bal}_n(R)$. If $M \in \mathrm{Im}(\lambda) \cap \mathrm{Ker}(\mu)$, then $M = \lambda(\mathbf{v})$ for some $\mathbf{v} \in \mathrm{Bal}_n(R)$. Then $0 = \mu(M) = \mu(\lambda(\mathbf{v})) = \mathbf{v}$, so that $M = 0$ and $\mathrm{Im}(\lambda) \cap \mathrm{Ker}(\mu) = 0$. Next for any $M \in \mathrm{Bal}_n(R)$, we have

$$\mu(M - (\lambda \circ \mu)(M)) = \mu(M) - \mu(M) = 0,$$

since $\mu \circ \lambda$ is the identity. Therefore $M - (\lambda \circ \mu)(M) \in \mathrm{Ker}(\mu)$ and $M \in \mathrm{Im}(\lambda) + \mathrm{Ker}(\mu)$; the result follows by definition of the direct sum.

2. This is a consequence of the definition of λ.

3. Let $M \in M_n(R)$ and define $S_1 = M + M^T$ and $S_2 = M - M^T$; then $2M = S_1 + S_2$. Notice that S_1 is symmetric and S_2 is skew symmetric. Hence $\mu(S_1) = 0$ and $\mu(S_2) = 2S_2 I$, from which it follows that $2\mu(M) = \mu(S_1) + \mu(S_2) = 2S_2 I$. Recall that R, being an ordered domain, has characteristic zero (see 2.1), and thus an equation $2A = 2B$ in $\mathrm{Bal}_n(R)$ implies that $A = B$. Therefore we may conclude that $\mu(M) = S_2 I$ and as a consequence that $\mu(M) = 0$ if and only if $S_2 I = 0$, i.e., S_2 is a skew-symmetric matrix with row sums equal to 0. $\qquad\square$

Transactions and digraphs

We end the chapter by describing yet another way of visualizing a transaction, this time geometrically by means of digraphs. First of all recall that a *directed graph*, or *digraph*, D consists of a non-empty set V and a binary relation E on V: thus

$$E \subseteq V \times V$$

and $u\,E\,v$ if and only if $(u,v) \in E$. The elements of V are called the
vertices of D and the elements (u,v) of E, written

$$\langle u,v \rangle,$$

are the *edges* of D. A geometric picture of the digraph is obtained
when the vertices are represented by points in the plane and an edge
$\langle u,v \rangle$ is represented by a directed line segment from u to v,

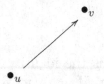

A *loop* in a digraph is an edge from a vertex to itself. Clearly a
digraph has no loops precisely when the corresponding relation E is
irreflexive, i.e., $u\,E\,u$ never holds. If there are edges $\langle u,v \rangle$ and $\langle v,u \rangle$,
these are called *parallel edges*. A digraph has no parallel edges if and
only if the corresponding relation E is antisymmetric, i.e., $u\,E\,v$ and
$v\,E\,u$ cannot both hold. The number of edges beginning at a vertex
v is·called the *out-degree* of v and the number of edges ending in v is
the *in-degree*. We shall be especially interested in digraphs in which
no vertex has positive in-degree and positive out-degree; notice that
a digraph with this property cannot have loops or parallel edges.

Now let us return to transactions. Suppose that $\mathbf{v} \in \mathrm{Bal}_n(R)$
where R is an ordered domain, and regard $\mathbf{v}$ as a transaction vector. We define a corresponding digraph D with vertex set the set
of accounts $\{a_1, a_2, \ldots, a_n\}$. An edge $\langle a_i, a_j \rangle$ is to be drawn from
account a_i to account a_j if $v_i < 0$ and $v_j > 0$, i.e., the transaction
debits a_j and credits a_i, while it may also affect other accounts.

It follows at once from the definition that the digraph of a transaction has the special property that no vertex can have positive in-
degree and positive out-degree. In fact a rather stronger property
holds.

(3.3.4). *If D is the digraph of a transaction vector, then the vertex
set of D is the union of three disjoint subsets V_0, V_1, V_2 such that
there is an edge from each vertex of V_1 to each vertex of V_2 and no
other edges are present in D.*

Proof
If D is the digraph of a transaction $\mathbf{v}$, define V_0, V_1, V_2 to be respectively the sets of accounts a_i for which $v_i = 0$, $v_i < 0$, $v_i > 0$. These sets have the property stated. $\square$

Example (3.3.4).
Consider the transaction vector

$$\begin{bmatrix} -550 \\ -500 \\ 200 \\ 0 \\ 850 \end{bmatrix}$$

The digraph of this balance vector is shown below, where for simplicity of notation the vertices have been labeled $1, \ldots, 5$:

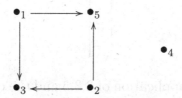

Here, for example, there is an edge from 1 to 5 since $v_1 < 0$ and $v_5 > 0$, and the vertex 4 is isolated in the digraph since the transaction does not affect account a_4. The three subsets of 3.3.5 are $V_0 = \{4\}$, $V_1 = \{1, 2\}$, $V_2 = \{3, 5\}$.

It is evident that the digraph of a balance vector depends only on the type of the balance vector, not on its actual entries. Indeed there is a bijective correspondence between types of balance vectors and digraphs with the property enunciated in 3.3.4.

(3.3.5). *There is a bijective function from the set of all types of balance vectors with n entries and the set of all digraphs D on a given n-element vertex set with the property that the vertex set of D is the union of three disjoint subsets V_0, V_1, V_2 such that there is an edge from each vertex of V_1 to each vertex of V_2 and no other edges are present in D.*

Proof
We have seen that every n-balance vector type determines a unique
digraph which has the stated property. To get a map in the other
direction we assume that D is a digraph with the property. The
balance vector type $\mathbf{t}$ corresponding to D is defined as follows. To
determine the i-component of $\mathbf{t}$ look at D; if the ith vertex v_i is
in V_1, so there is an edge from vertex i to some other vertex, then
$t_i = -$; if $v_i \in V_2$, there is an edge from some vertex to the ith
vertex and $t_i = +$; if $v_i \in V_0$, so that v_i is isolated, then $t_i = 0$. This
process defines a unique type vector since V_0, V_1, V_2 are disjoint sets.
Clearly these two maps are mutually inverse, so we have a bijection.

$\square$

For example, consider Example 3.3.4 once again. We see directly
from the digraph that the corresponding type vector is

$$\begin{bmatrix} - \\ - \\ + \\ 0 \\ + \end{bmatrix}.$$

As an immediate application of 3.3.5 and the count of transaction
types in 3.2.1, we obtain combinatorial information about digraphs
with the property described in 3.3.5.

(3.3.6). *The number of digraphs D with a fixed set of n vertices such
that the vertex set is the union of three disjoint subsets V_0, V_1, V_2 and
there is an edge from each vertex of V_1 to each vertex of V_2 while no
other edges are present in D is equal to $3^n - 2^{n+1} + 2$.*

Chapter Four

Abstract Accounting Systems

4.1. Allowable Transactions and Balances

In the last two chapters it has been shown how one can represent the state of an accounting system by a balance vector and a change in the state of the system by a transaction vector, which is itself a balance vector. Now an essential component of any accounting system is the set of rules by which the system operates. The next objective is to complete the definition of our basic model of an accounting system by specifying in algebraic form the rules that govern the operation of the system.

In any real life accounting system there will be certain transactions that would be regarded as improper. A transaction might be contrary to sound business practice or it might violate government regulations. For example, a transaction that leads to a transfer of funds from an employee's pension account to cash would not be permitted under normal circumstances. To exclude such undesirable operations, an accounting system should come equipped with a list of transactions that are regarded as valid operations for the system. These will be called *allowable transactions*.

Another feature of an accounting system is that, even if a transaction is allowable, its application might still be rejected if it caused an unacceptable balance to appear in some account. For example, in the case of a retail firm customer credit accounts are likely to have limits. A purchase on credit by a customer would not be permitted if it led to a balance that exceeded the customer's credit limit. There may also be minimum balances for certain reserve accounts in an accounting system. Thus one recognizes the existence of *allowable balances*, as well as allowable transactions.

Before a transaction is accepted by an accounting system, it must first be screened for allowability. Should it pass this test, the balance vector which results when the transaction is applied must be computed. If the new balance vector is allowable, the transaction is approved and applied to the system.

The discussion so far suggests that the basic model of an accounting system should include the following:

1. a set of accounts in a specified order;

2. a set of allowable transactions;

3. a set of allowable balance vectors.

Once an accounting system has been defined in this way, it is natural to regard it as an automaton, and this point of view will be explored in detail in Chapter 6. This in turn leads by well known procedures to such algebraic structures as monoids and groups. Equally important are algebraic concepts such as substructures and quotient structures of a specific structure, which can be applied to accounting systems to provide algebraic representations of standard accounting procedures. In this and subsequent chapters the pay-off for introducing abstract algebra into accounting theory will become apparent.

4.2. Defining an Accounting System

Our objective in this section is to formulate precisely the definition of an accounting system on n accounts over an ordered domain R. The first component of the definition is an n-element set A called the *set of accounts*. Next we introduce the notion of a *balance function* from A to R: this is a function

$$\alpha : A \to R$$

such that

$$\sum_{a \in A} \alpha(a) = 0_R.$$

There is a connection with balance vectors here, which will be explained shortly. To complete the specification of the model, we choose two sets of balance functions from A to R, say

$$T \quad \text{and} \quad B,$$

where B must not be empty. Then the triple

$$\mathcal{A} = (A|\ T|\ B)$$

is called *an abstract accounting system* on A over R, with the sets T and B determining respectively the transactions which may be applied and the account balances which may arise, in a manner which will be described.

In order to introduce balance vectors we first linearly order the account set A in some fixed way: let this be

$$\{a_1, a_2, \ldots, a_n\}.$$

If $\alpha : A \rightarrow R$ is a balance function, then α is completely determined by the balance vector

$$\begin{bmatrix} \alpha(a_1) \\ \alpha(a_2) \\ \vdots \\ \alpha(a_n) \end{bmatrix}.$$

Thus we can replace each balance function by its associated balance vector, so that T and B may be regarded as sets of n-balance vectors over R, i.e., as subsets of $\mathrm{Bal}_n(R)$. We call T and B the sets of *allowable transactions* and *allowable balances* of $\mathcal{A}$. But note that strictly speaking T consists of balance vectors $\mathbf{v}$ rather than the associated transactions $\tau_{\mathbf{v}}$.

The mode of operation of the accounting system will now be described. The system has an initial balance vector $\mathbf{b}(0) \in B$. A sequence of allowable transactions $\mathbf{v}(1), \mathbf{v}(2), \ldots, \mathbf{v}(m)$ is applied to the system, producing successive allowable balance vectors $\mathbf{b}(1), \mathbf{b}(2), \ldots, \mathbf{b}(m)$ where

$$\mathbf{b}(i + 1) = \mathbf{b}(i) + \mathbf{v}(i),$$

provided that $\mathbf{b}(i + 1)$ is allowable, i.e., it belongs to B: if this is not the case, then $\mathbf{b}(i + 1) = \mathbf{b}(i)$.

In practice the allowable transactions will be of two sorts. There may be specific allowable transactions with fixed entries, for example, fixed rent or mortgage payments. Then there may be entire types of transactions that are allowable: a transaction in a retail firm which debits cash and credits inventory and profit/loss would be of this type. It is therefore reasonable to replace the set T by two sets

$$T_0 \quad \text{and} \quad T_1$$

and write
$$\mathcal{A} = (A|\ T_0, T_1|\ B)$$
where T_0 is the list of allowable transaction types and T_1 is the list
of specific allowable transactions. The understanding here is that
every transaction of a given allowable type is allowable.

It is reasonable to regard the identity transaction, i.e., the zero
transaction vector, as allowable since it has no effect on balances of
the accounting system; we will assume henceforth without further
comment that *this transaction is allowable in all accounting systems.*

Accounting systems with one account are uninteresting: the balance is always zero. Systems with two accounts are scarcely more
interesting: the two accounts have balances that are negatives of
each other. The simplest interesting accounting system has three accounts. Such a system could represent the uncomplicated financial
position of a company or an individual with few assets or liabilities,
with one account representing total assets, one total liabilities and
an account giving the net worth. This simple system also represents
the most primitive type of financial report, wherein total balances
are given for all the asset, liability and equity accounts.

The digraph of an accounting system

A useful way of visualizing the operation of an accounting system
is by means of a *digraph* (or directed graph). Consider an accounting
system
$$\mathcal{A} = (A|\ T|\ B)$$
on n accounts $a_1, a_2, \ldots, a_n$. The digraph of $\mathcal{A}$ has vertex set
$$A\ =\ \{a_1, a_2, \ldots, a_n\},$$
and an edge
$$\langle a_j, a_i \rangle$$
is drawn from a_j to a_i if some allowable transaction has its ith entry
positive and its jth entry negative. Notice that the direction of the
edge is from negative to positive: thus an edge $\langle a_j, a_i \rangle$ indicates
a potential flow of value from account a_j to account a_i. It is often
convenient to label the vertex set by the integers $\{1, 2, \ldots, n\}$ rather
than the accounts. Note also the connection with the digraph of a
transaction defined in Chapter 3: the digraph of the system is just
the union of the digraphs of all the allowable transactions.

Example (4.2.1).

Consider a system over $\mathbb{Z}$ with five accounts, one specific allowable transaction and three allowable transaction types,

$$
\begin{bmatrix} - \\ 0 \\ 0 \\ 0 \\ + \end{bmatrix}, \quad
\begin{bmatrix} 0 \\ 0 \\ 0 \\ - \\ + \end{bmatrix}, \quad
\begin{bmatrix} 0 \\ + \\ + \\ - \\ 0 \end{bmatrix}, \quad
\begin{bmatrix} 0 \\ 200 \\ 100 \\ 0 \\ -300 \end{bmatrix}.
$$

The digraph of this accounting system is:

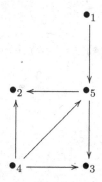

Sometimes one is only interested in the *undirected graph* of an accounting system $\mathcal{A}$, which arises when all the arrows in the digraph are omitted. If the graph of $\mathcal{A}$ is connected, i.e., there is a path between any two vertices, then $\mathcal{A}$ is termed a *connected accounting system*. Otherwise $\mathcal{A}$ is *disconnected*, in which case the graph decomposes into disjoint *connected components*. Clearly the system in Example 1 is connected. The undirected graph of an accounting system will be important when we consider the question of decomposability of accounting systems in 4.3.

Example (4.2.2).

Consider a 6-account system with allowable transactions and types

$$
\begin{bmatrix} - \\ + \\ 0 \\ 0 \\ 0 \\ 0 \end{bmatrix} , \quad
\begin{bmatrix} + \\ 0 \\ + \\ - \\ 0 \\ 0 \end{bmatrix} , \quad
\begin{bmatrix} 0 \\ 0 \\ 0 \\ 0 \\ + \\ - \end{bmatrix} , \quad
\begin{bmatrix} -100 \\ 200 \\ -100 \\ 0 \\ 0 \\ 0 \end{bmatrix} , \quad
\begin{bmatrix} 0 \\ -100 \\ 50 \\ 50 \\ 0 \\ 0 \end{bmatrix} .
$$

Here the graph is disconnected, with two connected components:

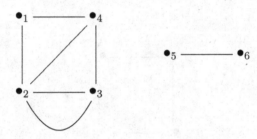

Thus the accounting system is disconnected.

It is clear from the definition that the digraph of an accounting system cannot contain loops. In fact this is the only restriction on the digraph.

(4.2.1). *Let D be a digraph with n vertices which has no loops. Then there is an accounting system on n accounts over any ordered domain R with digraph D.*

Proof
Let $1, 2, \ldots, n$ denote the vertices of D. The accounting system to be constructed has account set $A = \{a_1, a_2, \ldots, a_n\}$. If there is an edge from vertex i to vertex j in D, put the elementary transaction $\mathbf{e}(j, i)$ in the set of allowable transactions T. Then $\mathcal{A} = (A|\, T|\, \mathrm{Bal}_n(R))$ is an accounting system whose digraph is D. $\square$

This result should be compared with the much stronger condition for a digraph to be the digraph of a transaction, which is given in 3.3.4.

Feasible transactions and the feasible digraph

In an accounting system there are likely to be transactions that can be executed by means of a sequence of allowable transactions, but which are not themselves allowable. Such transactions do not contribute edges to the digraph of the system, but they can be used to augment it to form a larger digraph.

Consider an accounting system

$$\mathcal{A} = (A|\ T|\ B)$$

and let D be its digraph. A transaction which is the composite of a finite sequence of allowable transactions is called a *feasible transaction* for $\mathcal{A}$. Allowable transactions are feasible, but the converse need not be true. Since $\tau_{\mathbf{v}} \circ \tau_{\mathbf{w}} = \tau_{\mathbf{v}+\mathbf{w}}$, a typical feasible transaction vector has the form

$$a_1 \mathbf{v}(1) + a_2 \mathbf{v}(2) + \cdots + a_k \mathbf{v}(k)$$

where $\mathbf{v}(1), \mathbf{v}(2), \ldots, \mathbf{v}(k)$ are allowable transaction vectors and the a_i are non-negative integers. Let us write

$$\overline{T}$$

for the set of all feasible transactions for $\mathcal{A}$. Of course $T \subseteq \overline{T}$. Then we can form a new accounting system

$$\overline{\mathcal{A}} = (A|\ \overline{T}|\ B)$$

in which the set of allowable transactions is $\overline{T}$. The *feasible digraph* $\overline{D}$ *of* $\mathcal{A}$ is defined to be the digraph of $\overline{\mathcal{A}}$. It is evident that D is a subdigraph of the digraph $\overline{D}$. Notice that $\overline{D}$ represents flows of value in the system due to the action of feasible transactions, i.e., of sequences of allowable transactions.

Example (4.2.3).

Consider the accounting system with three accounts and two specific allowable transactions:

$$\mathbf{u} = \begin{bmatrix} -1 \\ 5 \\ -4 \end{bmatrix} \quad \text{and} \quad \mathbf{v} = \begin{bmatrix} 1 \\ -2 \\ 1 \end{bmatrix}.$$

The digraph of this system is

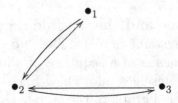

To find the feasible digraph we must identify the feasible trans-actions: these are all of the form

$$a\mathbf{u} + b\mathbf{v} = \begin{bmatrix} -a + b \\ 5a - 2b \\ -4a + b \end{bmatrix},$$

where a, b are non-negative integers. Now the inequalities $-a+b < 0$ and $-4a + b > 0$ are clearly contradictory. Therefore there cannot be an edge $\langle 1, 3 \rangle$ in the feasible digraph. On the other hand, setting $a = 1$, $b = 2$ yields the feasible transaction vector

$$\begin{bmatrix} 1 \\ 1 \\ -2 \end{bmatrix},$$

which demonstrates that there is an edge $\langle 3, 1 \rangle$. It follows that the feasible digraph of the system is:

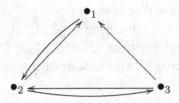

There are clearly limits to the amount of information about an accounting system that can be gleaned from its digraphs. For example, the digraphs do not enable us to tell what the allowable transactions are, but only the flows which they produce. Nor do they give information about the allowable balances. Nevertheless digraphs are a useful way of visualizing the effect of an allowable transaction or feasible transaction on an accounting system.

Equivalent accounting systems

Consider two accounting systems with the same set of accounts and over the same ordered domain,

$$\mathcal{A} = (A|\ T|\ B) \quad \text{and} \quad \mathcal{A}' = (A|\ T'|\ B').$$

Then $\mathcal{A}$ and $\mathcal{A}'$ are said to be *equivalent* if they have the same sets of feasible transactions: plainly this amounts to saying that an allowable transaction of one system is feasible in the other. Thus equivalent systems have the same feasible digraphs. What this means in practice is that, provided that balance restrictions are ignored, the two systems will arrive at the same final balance vector if they start from a common initial vector, albeit by means of different sequences of transactions. On the other hand, equivalent systems may have quite different sets of allowable transactions. Thus we should think of equivalent systems as possibly different systems having the same capacity to compute account balances.

Bounded accounting systems

A natural way to restrict the account balances in an accounting system is to place upper or lower limits on them. For example, a customer's account with a retail firm is likely to have a credit limit, which will appear as an upper bound after allowing for the positive sign of the account entry. A cash account might have a minimum balance, which would mean that there is an lower bound for the account balance. There might also be accounts without balance restrictions.

We proceed now to formalize these ideas for an accounting system $\mathcal{A} = (A|\ T|\ B)$ with n accounts over an ordered domain R. To allow unbounded values for some accounts, it is convenient to introduce the symbols $+\infty$ and $-\infty$ with their usual meanings. Thus the inequalities $-\infty < r < +\infty$ are valid for all $r \in R$.

A pair of functions

$$\lambda : A \to R \cup \{-\infty, \infty\} \quad \text{and} \quad \Lambda : A \to R \cup \{-\infty, \infty\}$$

is called a *bounding pair* provided that

$$\lambda(a_i) \leq \Lambda(a_i)$$

for $i = 1, 2, \ldots, n$. Next define

$$B(\lambda, \Lambda) = \{ \mathbf{v} \in \mathrm{Bal}_n(R) \mid \lambda(a_i) \leq v_i \leq \Lambda(a_i), \ i = 1, 2, \ldots, n \},$$

where an inequality says nothing and is to be ignored if $\lambda(a_i) = -\infty$ or $\Lambda(a_i) = +\infty$. Now form the accounting system

$$\mathcal{A} = (A|\ T|\ B|\ \lambda, \Lambda)$$

where $B \subseteq B(\lambda, \Lambda)$. For a balance vector to be allowable, it must fall in the set $B(\lambda, \Lambda)$, i.e., have its balances restricted by the functions λ, Λ; of course there might be further restrictions, so B could be a proper subset of $B(\lambda, \Lambda)$.

An accounting system which comes equipped with a bounding pair of functions (λ, Λ) will be called a *bounded accounting system*. If λ, Λ have all their values finite, i.e., not $\pm\infty$, then the system is termed *absolutely bounded*.

On the other hand, if all balance vectors are allowable, so that

$$\mathcal{A} = (A|\ T|\ \mathrm{Bal}_n(R)),$$

we call the system *unbounded*. Finally, if all balance vectors and all transaction vectors are allowable, so that

$$\mathcal{A} = (A|\ \mathrm{Bal}_n(R)|\ \mathrm{Bal}_n(R)),$$

then $\mathcal{A}$ is called a *free* accounting system.

4.3. Subaccounting Systems

In the study of many algebraic structures there are common concepts that appear at an early stage in the development of the theory. One such concept is the notion of a sub-structure, which means, roughly speaking, a structure that is contained inside a larger structure of the same type. Examples which come to mind include subspace and submodule. In view of this phenomenon it is reasonable to introduce the concept of a subaccounting system in accounting theory.

In order to come up with the "right" definition, we need to look at a real life system. In the case of a large firm there are likely to be subdivisions or units with a considerable degree of autonomy. Such a unit might have a set of accounts under its control and be able to execute transactions on these accounts, although such trans-actions would still need approval at senior management level. The unit's allowable transactions would likely not affect accounts which are outside its control. In addition, allowable balances for the unit would always have to be compatible with those for the entire system. These observations suggests how a subaccounting system should be defined.

Definition

Consider an accounting system with n accounts over an ordered domain R

$$\mathcal{A} = (A|\ T|\ B),$$

with the usual notation and conventions. An accounting system $\mathcal{A}' = (A'|\ T'|\ B')$ over R is said to be a *subaccounting system* of $\mathcal{A}$ if the following conditions are satisfied:

1. $A' \subseteq A$;

2. if $\mathbf{v} \in T \cup B$, then the restriction $\mathbf{v}|_{A'}$ of $\mathbf{v}$ to A' is a balance vector;

3. $T' = \{\mathbf{v}|_{A'} \mid \mathbf{v} \in T,\ \text{sppt}(\mathbf{v}) \subseteq A'\}$;

4. $B' = \{\mathbf{b}|_{A'} \mid \mathbf{b} \in B\}$.

Some explanation of these conditions is called for at this point, but first recall that the *support* of $\mathbf{v}$ is the set of accounts for which $\mathbf{v}$ has a non-zero entry,

$$\text{sppt}(\mathbf{v}) = \{a_i \mid v_i \neq 0\}.$$

Of course (1) asserts that each account of $\mathcal{A}'$ is an account of $\mathcal{A}$. According to (2) the restriction of an allowable vector of $\mathcal{A}$ to A' must be a balance vector; this ensures that the system $\mathcal{A}'$ is always in balance when transactions are applied to $\mathcal{A}$. The effect of (3) is that the allowable transactions of $\mathcal{A}'$ are the restrictions to A' of allowable transactions of $\mathcal{A}$ which do not affect accounts outside A'. Finally, (4) asserts that the allowable balance vectors of $\mathcal{A}'$ are precisely the restrictions to A' of allowable balance vectors of $\mathcal{A}$.

It is obvious that every accounting system is a subsystem of itself. A subsystem of a system $\mathcal{A}$ with fewer accounts than $\mathcal{A}$ is called a *proper subsystem* of $\mathcal{A}$. In general an accounting system might have no proper subsystems; in fact it possible to give a criterion for the existence of proper subsystems.

(4.3.1). *An accounting system $\mathcal{A} = (A|\ T|\ B)$ has a proper subsystem if and only if there is a proper non-empty subset A' of A such that $\mathbf{v}|_{A'}$ is a balance vector whenever $\mathbf{v} \in T \cup B$.*

Proof
If $\mathcal{A}' = (A'|\ T'|\ B')$ is a proper subsystem of $\mathcal{A}$, then $A' \neq A$ and A' has the property stated by part (2) of the definition. Conversely, let A' be a subset of A satisfying the condition and define

$$T' = \{\mathbf{v}|_{A'} \mid \mathbf{v} \in T,\ \text{sppt}(\mathbf{v}) \subseteq A'\}$$

and
$$B' = \{\mathbf{b}|_{A'} \mid \mathbf{b} \in B\}.$$

Let $\mathcal{A}'$ be the accounting system $(A'|\ T'|\ B')$. Then $\mathcal{A}'$ is a proper subsystem of $\mathcal{A}$ since properties (1)–(4) of the definition are valid.

$\square$

The concept of a subsystem is illustrated by some examples.

Example (4.3.1).

Consider the accounting system $\mathcal{A}$ over $\mathbb{Z}$ on accounts a_1, a_2, a_3, a_4 with allowable transactions

$$\begin{bmatrix} 50 \\ 0 \\ -50 \\ 0 \end{bmatrix},\ \begin{bmatrix} 0 \\ -50 \\ 0 \\ 50 \end{bmatrix},$$

whose the allowable balance vectors are all vectors of the form

$$\begin{bmatrix} x \\ y \\ -x \\ -y \end{bmatrix},\ 0 \leq x, y \leq 200.$$

This has a proper subsystem on accounts a_1, a_3 with one allowable transaction vector $\begin{bmatrix} 50 \\ -50 \end{bmatrix}$ and allowable balance vectors $\begin{bmatrix} x \\ -x \end{bmatrix}$, where $0 \leq x \leq 200$.

Example (4.3.2).

Let $\mathcal{A}$ be the accounting system over $\mathbb{Z}$ on accounts a_1, a_2, a_3, a_4 with allowable transactions

$$\begin{bmatrix} 50 \\ 0 \\ -50 \\ 0 \end{bmatrix}, \quad \begin{bmatrix} 0 \\ -50 \\ 0 \\ 50 \end{bmatrix}, \quad \begin{bmatrix} -50 \\ 50 \\ 50 \\ -50 \end{bmatrix},$$

and allowable balance vectors

$$\begin{bmatrix} x \\ -x \\ y \\ -y \end{bmatrix}, \quad 0 \le x \le 200.$$

This accounting system has no proper subsystems. For if A' were the account set of a proper subsystem, we see from the allowable transactions that the only possibilities for A' would be $\{a_1, a_3\}$ and $\{a_2, a_4\}$. However, in each case the restriction to A' of an allowable balance vector need not be a balance vector, so neither is possible.

Joins of accounting systems

Our next object is to describe a method for splicing together a number of different accounting systems to produce a larger system, called the *join*, which contains the original systems as subsystems. In algebra this is a familiar procedure, a typical example being the direct sum of vector spaces. In fact this could occur in real life accounting situations, for example, when two or more previously independent divisions of a company are consolidated into a single entity. Of course in such a case this might result in duplicate accounts, for example if each division had an account with the same third party. Thus the join operation would have to be followed by an amalgamation of accounts, a procedure that will be made precise in Chapter 5 by introducing quotient systems. Thus the consolidation process can be visualized as a join followed by passage to a quotient system. Taking the opposite point of view one might seek to break up an accounting system by expressing it as a join of smaller subsystems.

Definition of the join

We begin with a set of k accounting systems over an ordered domain R,

$$\mathcal{A}^{(i)} = (A^{(i)} | T^{(i)} | B^{(i)}), \quad i = 1, 2, \ldots, k,$$

where the account sets $A^{(1)}, A^{(2)}, \ldots, A^{(k)}$ are assumed to be mutually disjoint. Our object is to construct a new accounting system

$$\mathcal{A} = \mathcal{A}^{(1)} \vee \mathcal{A}^{(2)} \vee \cdots \vee \mathcal{A}^{(k)},$$

called the *join* of the $\mathcal{A}^{(i)}$, which has each $\mathcal{A}^{(i)}$ as a subsystem. Our first move is to specify the account set for $\mathcal{A}$: this will be the union

$$A = \bigcup_{i=1}^{k} A^{(i)}.$$

Let the accounts in A be linearly ordered first by the order of sets

$$A^{(1)}, A^{(2)}, \ldots, A^{(k)},$$

and then by using the order of elements within each set $A^{(i)}$.

Writing n_i for $|A^{(i)}|$, the number of accounts in $\mathcal{A}^{(i)}$, we see that the number of accounts in $\mathcal{A}$ is

$$n = \sum_{i=1}^{k} n_i.$$

If $\mathbf{v} \in \mathrm{Bal}_{n_i}(R)$, define a vector

$$\mathbf{v}^* \in \mathrm{Bal}_n(R)$$

by the following rule:

$$v_j^* = \begin{cases} v_j & \text{if } \sum_{r=1}^{i-1} n_r < j \le \sum_{r=1}^{i} n_r \\ 0 & \text{otherwise} \end{cases}$$

Thus, in passing from $\mathbf{v}$ to $\mathbf{v}^*$, we insert zeros in $\mathbf{v}$ for all accounts in $A^{(j)}$ where $j \ne i$, noting that $\mathbf{v}^*$ is also a balance vector. (The same rule may be applied with a type in place of $\mathbf{v}$).

The set of allowable transactions and transaction types for $\mathcal{A}$ is defined to be

$$T = \left\{ \mathbf{v}^* | \mathbf{v} \in T^{(i)}, \quad i = 1, 2, \ldots, k \right\}.$$

Thus the allowable transactions for $\mathcal{A}$ arise from those of the original systems $\mathcal{A}^{(i)}$ by inserting zeros at appropriate points in the column vectors.

The procedure for defining the set of allowable balances is a little different since it is natural to impose the balances restrictions of $\mathcal{A}^{(i)}$ only on the accounts in $A^{(i)}$. Therefore we define the set of allowable balance vectors for $\mathcal{A}$ to be

$$B = \{\mathbf{b}(1)^* + \mathbf{b}(2)^* + \cdots + \mathbf{b}(k)^* \mid \mathbf{b}(i) \in B^{(i)}, \ i = 1, 2, \dots, k\}.$$

The point to observe here is that $\mathbf{b}(1)^* + \mathbf{b}(2)^* + \cdots + \mathbf{b}(k)^*$ affects balances of accounts in $A^{(i)}$ only through its $A^{(i)}$-component $\mathbf{b}(i)$. Finally, the *join* $\mathcal{A} = \mathcal{A}^{(1)} \vee \mathcal{A}^{(2)} \vee \cdots \vee \mathcal{A}^{(k)}$ of the systems $\mathcal{A}^{(i)}$ is defined to be

$$\mathcal{A} = (A\mid T\mid B).$$

A basic property of join is stated next.

(4.3.2). *In a join of accounting systems* $\mathcal{A} = \mathcal{A}^{(1)} \vee \mathcal{A}^{(2)} \vee \cdots \vee \mathcal{A}^{(k)}$ *each* $\mathcal{A}^{(i)}$ *is a subaccounting system.*

Proof
Let $\mathcal{A}^{(i)} = (A^{(i)}\mid T^{(i)}\mid B^{(i)})$. A typical allowable transaction vector of $\mathcal{A}$ has the form $\mathbf{v}^*$ where $\mathbf{v} \in T^{(i)}$ for some i, and $\mathbf{v}^*|_{A^{(i)}} = \mathbf{v}$; also $\mathrm{sppt}(\mathbf{v}^*) \subseteq A^{(i)}$ since $\mathbf{v}^*$ has zero entries for accounts not in $A^{(i)}$. A typical allowable balance vector for $\mathcal{A}$ has the form $\mathbf{b} = \mathbf{b}(1)^* + \cdots + \mathbf{b}(k)^*$ where $\mathbf{b}(i) \in B^{(i)}$, and clearly $\mathbf{b}|_{A^{(i)}} = \mathbf{b}(i)$. Hence $\mathcal{A}^{(i)}$ is a subsystem of $\mathcal{A}$ □

Example (4.3.3). To illustrate the join procedure consider the join of two accounting systems $\mathcal{A}$ and $\mathcal{A}'$ defined as follows.

$$\mathcal{A} = \left(\{a_1, a_2, a_3\} \ \left|\begin{bmatrix} - \\ + \\ + \end{bmatrix}, \begin{bmatrix} 0 \\ 50 \\ -50 \end{bmatrix}, \begin{bmatrix} 200 \\ -150 \\ -50 \end{bmatrix}\right|\ B\right)$$

and

$$\mathcal{A}' = \left(\{a_1', a_2', a_3', a_4'\} \ \left|\begin{bmatrix} - \\ + \\ - \\ - \end{bmatrix}, \begin{bmatrix} 0 \\ 100 \\ 0 \\ -100 \end{bmatrix}\right|\ B'\right)$$

where B and B' are subsets of $\mathrm{Bal}_3(\mathbb{Z})$ and $\mathrm{Bal}_4(\mathbb{Z})$ consisting of vectors all of whose entries lie in the respective finite intervals $[m, M]$ and $[m', M']$. Thus $\mathcal{A}$ and $\mathcal{A}'$ are absolutely bounded systems.

The join $\bar{\mathcal{A}} = \mathcal{A} \vee \mathcal{A}'$ has seven accounts $a_1, a_2, a_3, a_1', a_2', a_3', a_4'$. The allowable transactions and types for $\bar{\mathcal{A}}$ arising from $\mathcal{A}$ are:

$$
\begin{bmatrix} 0 \\ 50 \\ -50 \end{bmatrix}^* = \begin{bmatrix} 0 \\ 50 \\ -50 \\ 0 \\ 0 \\ 0 \\ 0 \end{bmatrix}, \quad
\begin{bmatrix} 200 \\ -150 \\ 50 \end{bmatrix}^* = \begin{bmatrix} 200 \\ -150 \\ -50 \\ 0 \\ 0 \\ 0 \\ 0 \end{bmatrix},
$$

and

$$
\begin{bmatrix} - \\ + \\ + \end{bmatrix}^* = \begin{bmatrix} - \\ + \\ + \\ 0 \\ 0 \\ 0 \\ 0 \end{bmatrix}.
$$

Also the allowable transactions coming from $\mathcal{A}'$ are

$$
\begin{bmatrix} 0 \\ 100 \\ 0 \\ -100 \end{bmatrix}^* = \begin{bmatrix} 0 \\ 0 \\ 0 \\ 0 \\ 100 \\ 0 \\ -100 \end{bmatrix}, \quad
\begin{bmatrix} - \\ + \\ - \\ - \end{bmatrix}^* = \begin{bmatrix} 0 \\ 0 \\ 0 \\ - \\ + \\ - \\ - \end{bmatrix}.
$$

The allowable balance vectors for $\mathcal{A}$ are of the form

$$
\begin{bmatrix} x \\ y \\ -x - y \end{bmatrix}
$$

where $x, y, -x-y$ lie in $[m, M]$, while those for $\mathcal{A}'$ are of the form

$$
\begin{bmatrix} x' \\ y' \\ z' \\ -x' - y' - z' \end{bmatrix}
$$

with $x', y', z', -x'-y'-z'$ in $[m', M']$. Thus a typical allowable balance vector for the join $\mathcal{A}^*$ is

$$
\begin{bmatrix} x \\ y \\ -x-y \end{bmatrix}^* + \begin{bmatrix} x' \\ y' \\ z' \\ -x'-y'-z' \end{bmatrix}^* = \begin{bmatrix} x \\ y \\ -x-y \\ 0 \\ 0 \\ 0 \\ 0 \end{bmatrix} + \begin{bmatrix} 0 \\ 0 \\ 0 \\ x' \\ y' \\ z' \\ -x'-y'-z' \end{bmatrix},
$$

which equals

$$
\begin{bmatrix} x \\ y \\ -x-y \\ x' \\ y' \\ z' \\ -x'-y'-z' \end{bmatrix},
$$

where $x, y, -x-y$ and $x', y', z', -x'+y'-z'$ are in $[m, M]$ and $[m', M']$ respectively.

A notable property of the join of accounting systems is that a transaction that can be executed by one of the factors of the join can also be executed by the system as a whole. More precisely the following is true.

(4.3.3). *Let* $\mathcal{A} = \mathcal{A}^{(1)} \vee \mathcal{A}^{(2)} \vee \cdots \vee \mathcal{A}^{(k)}$ *be a join of accounting systems, where* $\mathcal{A}^{(i)} = (A^{(i)}|T^{(i)}|B^{(i)})$. *Suppose that* $\mathbf{v}(i) \in T^{(i)}$ *and* $\mathbf{b}(i), \mathbf{b}(i)+\mathbf{v}(i) \in B^{(i)}$ *for a fixed* i. *Then there exist* $\mathbf{v} \in T$ *and* $\mathbf{b} \in B$ *such that* $\mathbf{b}+\mathbf{v} \in B$, $\mathbf{b}|_{A^{(i)}} = \mathbf{b}(i)$ *and* $(\mathbf{b}+\mathbf{v})|_{A^{(i)}} = \mathbf{b}(i) + \mathbf{v}(i)$.

Proof
Choose any $\mathbf{b}(j) \in B^{(j)}$ for $j \neq i$. Define $\mathbf{b} = \sum_{j=1}^{k} \mathbf{b}(j)^*$ and put $\mathbf{v} = \mathbf{v}(i)^*$; thus $\mathbf{b} \in B$ and $\mathbf{v} \in T$. Also $(\mathbf{b}+\mathbf{v})|_{A^{(j)}} = \mathbf{b}(j) + \mathbf{v}|_{A^{(j)}}$. Since $\mathbf{v}|_{A^{(j)}}$ equals $\mathbf{v}(i)$ if $j = i$ and $\mathbf{0}$ if $j \neq i$, we have $\mathbf{b} + \mathbf{v} \in B$. Hence $\mathbf{b}$ and $\mathbf{v}$ have the required properties. $\square$

However, the property of 4.3.3 does not hold for arbitrary subsystems: there can be transactions executable in a subsystem which are not executable in the whole accounting system.

Example (4.3.4).

Let $\mathcal{A}$ be the accounting system with accounts a_1, a_2, a_3, a_4, allowable transactions

$$\begin{bmatrix} 50 \\ -50 \\ 0 \\ 0 \end{bmatrix}, \begin{bmatrix} 0 \\ 0 \\ 50 \\ -50 \end{bmatrix}$$

and allowable balances

$$\begin{bmatrix} 100 \\ -100 \\ 200 \\ -200 \end{bmatrix}, \begin{bmatrix} 100 \\ -100 \\ 250 \\ -250 \end{bmatrix}, \begin{bmatrix} 150 \\ -150 \\ 350 \\ -350 \end{bmatrix}.$$

The system $\mathcal{A}$ has a subsystem $\mathcal{A}'$ on accounts a_1, a_2 with allowable transaction $\begin{bmatrix} 50 \\ -50 \end{bmatrix}$ and allowable balances $\begin{bmatrix} 100 \\ -100 \end{bmatrix}$ and $\begin{bmatrix} 150 \\ -150 \end{bmatrix}$.

Suppose that $\mathcal{A}'$ has initial balance $\begin{bmatrix} 100 \\ -100 \end{bmatrix}$ and the allowable transaction is applied; then this results in the allowable balance $\begin{bmatrix} 150 \\ -150 \end{bmatrix}$. However, this transaction cannot be executed in $\mathcal{A}$. For if it could, the initial balance vector would have to be one of the vectors

$$\begin{bmatrix} 100 \\ -100 \\ 200 \\ -200 \end{bmatrix}, \begin{bmatrix} 100 \\ -100 \\ 250 \\ -250 \end{bmatrix};$$

on the other hand, the final balance vector after applying an allowable transaction must be

$$\begin{bmatrix} 150 \\ -150 \\ 350 \\ -350 \end{bmatrix}.$$

However, neither transaction gives the correct answer.

Decomposable accounting systems

For the remainder of the chapter we shall study the question of when an accounting system can be expressed as a join of proper subsystems. First some terminology. If an accounting system $\mathcal{A}$ can be written as the join of two or more subsystems, then it will be called *decomposable*: note that the subsystems in question will necessarily be proper. If this is not possible, then $\mathcal{A}$ is said to be *indecomposable*. It is natural to ask how one can tell if a given system is decomposable. The first result provides a necessary condition for decomposability.

(4.3.4). *Let $\mathcal{A}$ be a decomposable accounting system with $\mathcal{A} = \mathcal{A}^{(1)} \vee \mathcal{A}^{(2)} \vee \cdots \vee \mathcal{A}^{(k)}$, $k \geq 2$. Then $\mathcal{A}$ is disconnected and each connected component is contained in some $A^{(i)}$ where $\mathcal{A}^{(i)} = (A^{(i)}|T^{(i)}|B^{(i)})$.*

Proof
Let G and $G^{(i)}$ be the respective graphs of $\mathcal{A}$ and $\mathcal{A}^{(i)}$. By definition of the join, the vertex set of G is the union of the (disjoint) vertex sets of the $G^{(i)}$, and also the edges of G are the edges of all the $G^{(i)}$. Thus G is the union of the disjoint subgraphs $G^{(i)}$, so G is a disconnected graph, and hence the system $\mathcal{A}$ is disconnected. Also each connected component of $\mathcal{A}$, i.e., of G, is contained in one of the subsets $A^{(i)}$. $\qquad\square$

One might hope that the converse of 4.3.4 would be true, but this is not the case, the reason being that disconnectedness does not place restrictions on the allowable balance vectors.

Example (4.3.5).

Let $\mathcal{A}$ be an accounting system with accounts a_1, a_2, a_3, a_4, allowable transactions

$$\begin{bmatrix} 50 \\ -50 \\ 0 \\ 0 \end{bmatrix} \text{ and } \begin{bmatrix} 0 \\ 0 \\ 100 \\ -100 \end{bmatrix},$$

and allowable balance vectors

$$\begin{bmatrix} x \\ y \\ z \\ t \end{bmatrix}$$

where $-100 \leq x, y, z, t \leq 200$ and $x + y + z + t = 0$. Clearly $\mathcal{A}$ is disconnected and its graph is:

However, $\mathcal{A}$ is indecomposable. Indeed, assume that $\mathcal{A} = \mathcal{A}^{(1)} \vee \mathcal{A}^{(2)}$ with $\mathcal{A}^{(i)} = (A^{(i)} \mid T^{(i)} \mid B^{(i)})$, $i = 1, 2$, and $A^{(i)} \neq A$. Then a_1 and a_2 belong to $A^{(1)}$, say, since a_1 and a_2 are affected by an allowable transaction of $\mathcal{A}$ and hence of $\mathcal{A}^{(1)}$ or $\mathcal{A}^{(2)}$. Clearly neither a_3 nor a_4 can be in $A^{(1)}$, otherwise $A^{(1)} = A$. Therefore $A^{(1)} = \{a_1, a_2\}$ and $A^{(2)} = \{a_3, a_4\}$. But if $\mathbf{b}$ is the allowable balance vector with components $50, 100, -100, -50$, then $\mathbf{b}|_{A^{(1)}}$ is not a balance vector. Therefore $\mathcal{A}$ cannot be decomposable.

A key tool in studying the decomposability of accounting systems is the support of a balance vector. Using this concept we can state a necessary and sufficient condition for an accounting system to be decomposable.

(4.3.5). *Let $\mathcal{A} = (A \mid T \mid B)$ be an accounting system over an ordered domain R. Then $\mathcal{A}$ is decomposable if and only if there is a partition $A = A^{(1)} \cup A^{(2)} \cup \cdots \cup A^{(k)}$ of A with $k > 1$ which has the following properties:*
  (a) if $\mathbf{v} \in T$, then the support of $\mathbf{v}$ is a subset of some $A^{(i)}$;
  (b) if $\mathbf{b} \in B$, then $\mathbf{b}|_{A^{(i)}}$ is a balance vector for $i = 1, 2, \ldots, k$;
  (c) if $\mathbf{b}(1), \mathbf{b}(2), \ldots, \mathbf{b}(k) \in B$, then

$$(\mathbf{b}(1)|_{A^{(1)}})^* + (\mathbf{b}(2)|_{A^{(2)}})^* + \cdots + (\mathbf{b}(k)|_{A^{(k)}})^* \in B.$$

Proof
Assume that $\mathcal{A}$ is decomposable and $\mathcal{A} = \mathcal{A}^{(1)} \vee \mathcal{A}^{(2)} \vee \cdots \vee \mathcal{A}^{(k)}$ where $k > 1$: write $\mathcal{A}^{(i)} = (A^{(i)} \mid T^{(i)} \mid B^{(i)})$. Then $A^{(1)} \cup A^{(2)} \cup \cdots \cup A^{(k)}$ is a proper partition of A. If $\mathbf{v} \in T$, then by definition the transaction $\tau_{\mathbf{v}}$ affects only accounts in some $A^{(i)}$ and thus $\mathrm{sppt}(\mathbf{v}) \subseteq A^{(i)}$. If $\mathbf{b} \in B$, then by definition $\mathbf{b}|_{A^{(i)}} \in B^{(i)}$, so that $\mathbf{b}|_{A^{(i)}}$ is a balance vector for each i. Finally (c) is valid by definition of the allowable balances in a join.

Conversely, assume that there is a partition $A = A^{(1)} \cup A^{(2)} \cup \cdots \cup A^{(k)}$ satisfying conditions (a), (b), (c). We show that $\mathcal{A}$ is decomposable. The idea is to define a subsystem $\mathcal{A}^{(i)}$ for $i = 1, 2, \ldots, k$, with account set $A^{(i)}$. If $\mathbf{v} \in \mathrm{Bal}_n(R)$, write $\mathbf{v}^{(i)} = \mathbf{v}|_{A^{(i)}}$. If $\mathbf{v} \in T$, then $\mathrm{sppt}(\mathbf{v}) \subseteq A_i$ for some i by (a) and hence $\mathbf{v}^{(i)}$ is a balance vector. Define

$$T^{(i)} = \{\mathbf{v}^{(i)} \mid \mathbf{v} \in T, \ \mathrm{sppt}(\mathbf{v}) \subseteq A_i\}.$$

If $\mathbf{b} \in B$, then $\mathbf{b}^{(i)}$ is a balance vector by (b); now define

$$B^{(i)} = \{\mathbf{b}^{(i)} \mid \mathbf{b} \in B\}.$$

Then $\mathcal{A}^{(i)} = (A^{(i)} \mid T^{(i)} \mid B^{(i)})$ is an accounting system. We will show that $\mathcal{A} = \mathcal{A}^{(1)} \vee \mathcal{A}^{(2)} \vee \cdots \vee \mathcal{A}^{(k)}$.

Firstly A is the union of the disjoint sets $A^{(i)}$. If $\mathbf{v} \in T$, then $\mathrm{sppt}(\mathbf{v}) \subseteq A^{(i)}$ for some i and $(\mathbf{v}^{(i)})^* = \mathbf{v}$; moreover $\mathbf{v}^{(i)}$ is a typical element of $T^{(i)}$. Also $\tau_{\mathbf{v}}$ affects only accounts in $A^{(i)}$. If $\mathbf{b} \in B$, then $\mathbf{b}^{(i)} \in B^{(i)}$ and $\mathbf{b} = \sum_{i=1}^{k}(\mathbf{b}^{(i)})^*$. Finally, if $\mathbf{b}(i) \in B^{(i)}$, then $\mathbf{b}(i) = \mathbf{b}(i)^*|_{A^{(i)}}$ and so we have

$$\sum_{i=1}^{k} \mathbf{b}(i)^* = \sum_{i=1}^{k}(\mathbf{b}(i)^*|_{A^{(i)}})^* \in B$$

by (c). This completes the proof that $\mathcal{A} = \mathcal{A}^{(1)} \vee \mathcal{A}^{(2)} \vee \cdots \vee \mathcal{A}^{(k)}$, and hence $\mathcal{A}$ is decomposable.

$\square$

Example (4.3.6).

Consider the 5-account system $\mathcal{A}$ which has allowable transactions

$$\begin{bmatrix} - \\ 0 \\ 0 \\ + \\ 0 \end{bmatrix}, \begin{bmatrix} - \\ 0 \\ 0 \\ + \\ - \end{bmatrix}, \begin{bmatrix} 0 \\ 50 \\ -50 \\ 0 \\ 0 \end{bmatrix},$$

and allowable balance vectors

$$\begin{bmatrix} x \\ y \\ -y \\ z \\ -x - z \end{bmatrix},$$

where $0 \le x, y, z \le 200$.

The supports of the allowable transactions are $\{a_1, a_4\}$, $\{a_2, a_3\}$, $\{a_1, a_4, a_5\}$ respectively. Each of these subsets lies inside a member of the partition $\{a_2, a_3\} \cup \{a_1, a_4, a_5\}$. The restrictions of an allowable balance vector to the subsets of this partition are balance vectors, and also condition (c) in 4.3.5 holds. Therefore $\mathcal{A}$ is decomposable.

In fact $\mathcal{A} = \mathcal{A}^{(1)} \vee \mathcal{A}^{(2)}$ where $\mathcal{A}^{(1)}, \mathcal{A}^{(2)}$ are defined as follows. First of all $\mathcal{A}^{(1)}$ has account set $\{a_1^{(1)}, a_2^{(1)}\}$, with $a_1^{(1)} = a_2$, $a_2^{(1)} = a_3$, and allowable transaction $\begin{bmatrix} 50 \\ -50 \end{bmatrix}$, and allowable balance vectors $\begin{bmatrix} y \\ -y \end{bmatrix}$, $0 \le y \le 200$. Then $\mathcal{A}^{(2)}$ has account set $\{a_1^{(2)}, a_2^{(2)}, a_3^{(2)}\}$ with $a_1^{(2)} = a_1$, $a_2^{(2)} = a_4$, $a_3^{(2)} = a_5$ and allowable transactions

$$\begin{bmatrix} - \\ + \\ 0 \end{bmatrix} \quad \text{and} \quad \begin{bmatrix} - \\ + \\ - \end{bmatrix},$$

and allowable balances

$$\begin{bmatrix} x \\ z \\ -x - z \end{bmatrix}, \quad 0 \le x, z \le 200.$$

Of course, the system $\mathcal{A}$ is disconnected, its graph being

Notice that in the last example the two subsystems $\mathcal{A}^{(1)}$ and $\mathcal{A}^{(2)}$ are indecomposable since their graphs are connected. It is an interesting fact that every accounting system can be expressed as the join of a set of indecomposable subsystems, a result which underscores the significance of indecomposable systems.

(4.3.6). *Let $\mathcal{A}$ be an arbitrary accounting system. Then $\mathcal{A}$ can be expressed in the form*

$$\mathcal{A} = \mathcal{A}^{(1)} \vee \mathcal{A}^{(2)} \vee \cdots \vee \mathcal{A}^{(k)},$$

where the $\mathcal{A}^{(i)}$ are indecomposable subsystems.

Proof

We establish the existence of the decomposition by induction on the number of accounts. If $\mathcal{A}$ is indecomposable, there is nothing to prove, so we assume it is decomposable. Thus $\mathcal{A} = \mathcal{B} \vee \mathcal{C}$ where $\mathcal{B}, \mathcal{C}$ are subsystems. Since $\mathcal{B}$ and $\mathcal{C}$ have fewer accounts than $\mathcal{A}$, each of them is a join of indecomposable subsystems. Therefore the same is true of $\mathcal{A}$ and the result is proved. $\qquad\square$

We conclude with an example where a given accounting system is expressed as a join of indecomposable subsystems.

Example (4.3.7).

Let $\mathcal{A}$ be the accounting system with accounts a_i, $i = 1, 2, \ldots, 8$, where the allowable transaction vectors are:

$$
\begin{bmatrix} 0 \\ -50 \\ 50 \\ 0 \\ 0 \\ 0 \\ 0 \\ 0 \end{bmatrix},
\begin{bmatrix} 0 \\ 0 \\ 0 \\ -50 \\ 50 \\ 0 \\ 0 \\ 0 \end{bmatrix},
\begin{bmatrix} -100 \\ 50 \\ 50 \\ 0 \\ 0 \\ 0 \\ 0 \\ 0 \end{bmatrix},
\begin{bmatrix} 0 \\ 0 \\ 0 \\ 0 \\ 0 \\ 100 \\ -100 \\ 0 \end{bmatrix},
\begin{bmatrix} 0 \\ 0 \\ 0 \\ 0 \\ 0 \\ 0 \\ -100 \\ 100 \end{bmatrix},
\begin{bmatrix} 0 \\ 0 \\ 0 \\ 0 \\ 0 \\ 100 \\ 0 \\ -100 \end{bmatrix}
$$

The allowable balance vectors for $\mathcal{A}$ are the vectors of the form

$$
\mathbf{b} = \begin{bmatrix} x \\ y \\ -x - y \\ z \\ -z \\ s \\ t \\ -s - t \end{bmatrix},
$$

where $0 \le x, y, z, s, t \le 500$, The graph of the system has three connected components $\{a_1, a_2, a_3\}, \{a_4, a_5\}, \{a_6, a_7, a_8\}$.

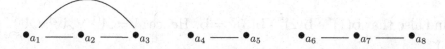

We look for indecomposable subsystems by examining the supports of the allowable transactions. In fact there are three obvious candidates. The first is $\mathcal{A}^{(1)}$ with accounts a_1, a_2, a_3 and allowable transaction vectors

$$\begin{bmatrix} 0 \\ -50 \\ 50 \end{bmatrix}, \begin{bmatrix} -100 \\ 50 \\ 50 \end{bmatrix}.$$

The second subsystem $\mathcal{A}^{(2)}$ has accounts a_4, a_5 and allowable transaction vector

$$\begin{bmatrix} -50 \\ 50 \end{bmatrix}.$$

The last subsystem $\mathcal{A}^{(3)}$ has accounts $\{a_6, a_7, a_8\}$ and allowable transactions

$$\begin{bmatrix} 100 \\ -100 \\ 0 \end{bmatrix}, \begin{bmatrix} 0 \\ -100 \\ 100 \end{bmatrix} \begin{bmatrix} 100 \\ 0 \\ -100 \end{bmatrix}.$$

The subsystems $\mathcal{A}^{(1)}, \mathcal{A}^{(2)}, \mathcal{A}^{(3)}$ are all indecomposable since their graphs are connected.

The respective allowable balance vectors for $\mathcal{A}^{(1)}, \mathcal{A}^{(2)}, \mathcal{A}^{(3)}$ are all the vectors of the forms

$$\mathbf{b}(1) = \begin{bmatrix} x \\ y \\ -x-y \end{bmatrix}, \mathbf{b}(2) = \begin{bmatrix} z \\ -z \end{bmatrix} \text{ and } \mathbf{b}(3) = \begin{bmatrix} s \\ t \\ -s-t \end{bmatrix}$$

where $0 \leq x, y, z, s, t \leq 500$. Notice that

$$\mathbf{b}(1)^* = \begin{bmatrix} x \\ y \\ -x-y \\ 0 \\ 0 \\ 0 \\ 0 \\ 0 \end{bmatrix}, \mathbf{b}(2)^* = \begin{bmatrix} 0 \\ 0 \\ 0 \\ z \\ -z \\ 0 \\ 0 \\ 0 \end{bmatrix}, \mathbf{b}(3)^* = \begin{bmatrix} 0 \\ 0 \\ 0 \\ 0 \\ 0 \\ s \\ t \\ -s-t \end{bmatrix},$$

and also that $\mathbf{b}(1)^* + \mathbf{b}(2)^* + \mathbf{b}(3)^* = \mathbf{b}$. Hence $\mathcal{A} = \mathcal{A}^{(1)} \vee \mathcal{A}^{(2)} \vee \mathcal{A}^{(3)}$.

Chapter Five

Quotient Systems and Homomorphisms

5.1. Introduction to the Quotient Concept

In our efforts to provide realistic models of accountancy systems it has been an on-going concern to show how standard operations in accounting can be represented by algebraic concepts. Thus balances, transactions and the rules of operation of an accounting system were seen to be representable within the framework of abstract accounting systems, as defined in Chapter 4. This program is continued in the present chapter by showing that the key algebraic concept of a *quotient structure* is able to model the procedure for generating reports on an accounting system.

A report on the financial condition of an organization consists of data obtained from the balance sheet by combining balances in groups of accounts which are subject to a common control mechanism. At the simplest level one could combine all the asset accounts, liability accounts and equity accounts in three new accounts. These combined accounts furnish limited but essential information about the state of the system. This is the most basic form of report.

More generally the accounts of an accounting system fall into a number of control groups, for example, accounts receivable, accounts payable. When the balances of all the accounts within the same group are combined, a report on the system is generated which provides information about the states of the various groups.

This procedure for generating a report is mirrored precisely by the algebraic notion of a quotient structure. By a process of combining accounts one arrives at a smaller accounting system called a quotient system. Thus we are motivated to introduce and analyze these systems.

An accounting system and its various quotient systems are linked by the procedure of forming reports. A more general question arises when one asks how two arbitrary accounting systems are related. Once again we can turn to abstract algebra for guidance. It is a very common question to ask what the relationship between two algebraic structures of the same type might be and how they can be compared. The usual approach is to look for functions between the structures which connect their internal operations: such functions are called *homomorphisms*.

It is our purpose here to introduce the concept of a homomorphism between the accounting systems and use this as a means to compare accounting systems. Homomorphisms are intimately related to quotient systems since the procedure for forming a report is an example of a homomorphism. Several other examples of homomorphisms that occur in practice are described below.

In the final section the theory of homomorphisms of accountancy systems is developed further, culminating in a series of "isomorphism theorems." These provide detailed information about homomorphisms and their relation to quotient systems; they also have applications to accounting systems. The appearance of such isomorphism theorems is a phenomenon which occurs throughout algebra.

5.2. Quotients of Accounting Systems

Consider an accounting system $\mathcal{A} = (A|\ T|\ B)$ over an ordered domain R. We will explain how to form new accounting systems from $\mathcal{A}$ called quotient systems. To form a quotient system of $\mathcal{A}$ we need to have an equivalence relation on the set of accounts A. First recall that an *equivalence relation* E on a set S is a binary relation for which the following properties are valid for all $a, b, c \in S$:

1. *reflexivity*, $a\ E\ a$ is always true;

2. *symmetry*, $a\ E\ b$ implies $b\ E\ a$;

3. *transitivity*, $a\ E\ b$ and $b\ E\ c$ imply that $a\ E\ c$.

The *equivalence class* containing a is the subset of S

$$[a] = [a]_E = \{b \mid a\ E\ b\}.$$

We recall the fundamental fact that *distinct equivalence classes are disjoint*. Thus the set A is the union of disjoint equivalence classes, i.e. the equivalence classes form a *partition* of A. Conversely, given a partition of a set A, an equivalence relation E on A is defined by declaring that two elements of A are E-equivalent if they belong to the same subset in the partition.

Returning to the accounting system $\mathcal{A} = (A|\ T|\ B)$, we choose an equivalence relation E on A and define

$$\overline{A} = \overline{A}_E = \{[a]_E \mid a \in A\},$$

i.e., $\overline{A}$ is the set of all distinct E-equivalence classes. Let $\overline{A}$ have $\overline{n} \leq n$ elements, say. The set $\overline{A}$ can be ordered by the smallest subscript appearing in each equivalence class.

Example (5.2.1).

Let $A = \{a_1, a_2, a_3, a_4, a_5\}$ and let E be the equivalence relation on A with associated partition

$$A = \{a_1\} \cup \{a_2, a_4, a_5\} \cup \{a_3\}.$$

The elements of $\overline{A}_E$ in order are

$$[a_1] = \{a_1\}, \quad [a_2] = \{a_2, a_4, a_5\}, \quad [a_3] = \{a_3\}.$$

In the general case we seek to define an accounting system on $\overline{A}$ called the quotient system of $\mathcal{A}$ by E. The next step is to specify the allowable transactions and balance vectors for the quotient system. Let $\mathbf{v} \in \mathrm{Bal}_n(R)$ and define a vector $\overline{\mathbf{v}} = \overline{\mathbf{v}}_E$ in $\mathrm{Bal}_{\overline{n}}(R)$ by the rule

$$\overline{v}_i = \sum_{a_j E a_i} v_j;$$

here the summation is over all j for which $a_j\ E\ a_i$. What this means is that the entries of $\mathbf{v}$ are totaled for accounts belonging to the same E-equivalence class. Observe that

$$\sum_{i=1}^{\overline{n}} \overline{v}_i = \sum_{i=1}^{n} v_i = 0,$$

so that $\overline{\mathbf{v}} \in \mathrm{Bal}_{\overline{n}}(R)$.

From this definition we quickly derive the rules

$$\overline{\mathbf{u} + \mathbf{v}} = \overline{\mathbf{u}} + \overline{\mathbf{v}} \quad \text{and} \quad \overline{r\mathbf{v}} = r\overline{\mathbf{v}},$$

where $\mathbf{u}, \mathbf{v} \in \text{Bal}_n(R)$ and $r \in R$. Hence the assignment $\mathbf{v} \mapsto \overline{\mathbf{v}}$ determines a homomorphism of R-modules from $\text{Bal}_n(R)$ to $\text{Bal}_{\overline{n}}(R)$.

Example (5.2.2).

Let E be the equivalence relation in Example 5.2.1, where there are five accounts and three subsets in the partition of accounts. Let

$$\mathbf{v} = \begin{bmatrix} v_1 \\ v_2 \\ v_3 \\ v_4 \\ -v_1 - v_2 - v_3 - v_4 \end{bmatrix} \in \text{Bal}_5(R).$$

Then, according to the definition,

$$\overline{\mathbf{v}} = \begin{bmatrix} v_1 \\ -v_1 - v_3 \\ v_3 \end{bmatrix} \in \text{Bal}_3(R).$$

After these preliminaries we are ready to formulate the definition of the quotient of the accounting system $\mathcal{A} = (A \mid T \mid B)$ determined by an equivalence relation E on A. Let

$$\overline{A} = \{[a_{i_j}]_E \mid j = 1, 2, \ldots, \overline{n}\}$$

be the set of distinct E-equivalence classes and define

$$\overline{T} = \{\overline{\mathbf{v}} \mid \mathbf{v} \in T\} \quad \text{and} \quad \overline{B} = \{\overline{\mathbf{v}} \mid \mathbf{v} \in B\}.$$

Then the *quotient system* of $\mathcal{A}$ by E is defined to be

$$\mathcal{A}/E = (\overline{A} \mid \overline{T} \mid \overline{B}).$$

Thus the accounts of $\mathcal{A}/E$ are the E-equivalence classes of accounts of $\mathcal{A}$, while the allowable transactions and balances of $\mathcal{A}/E$ arise from the corresponding entities of $\mathcal{A}$ by application of the function $\mathbf{v} \mapsto \overline{\mathbf{v}}$, i.e., by summing vector entries over each equivalence class.

Examples of quotient systems in accounting

(I) Reports

A quotient system of any accounting system $\mathcal{A} = (A|\ T|\ B)$ arises whenever one has an equivalence relation E, i.e. a partition of A. From the point of view of accounting, this happens when the account set is split up into various control groups and the quotient system represents a report on the groups.

The simplest case arises from the partition

$$A = A_e \cup A_a \cup A_l$$

where A_e, A_a, A_l are the respective sets of equity accounts, asset accounts, liability accounts. The resulting quotient system $\mathcal{A}/E$ has three accounts, namely total equity, total assets, total liabilities. This quotient provides the most basic type of report.

Another quotient system that might occur in practice arises from the partition

$$A = A_e \cup A_{ca} \cup A_{pa} \cup A_l,$$

where the four subsets of accounts represent equity accounts, current assets, plant assets, liabilities respectively: the quotient system provides a report on these control groups.

Even within asset accounts we can consider other control groups, such as fixed assets, inventories, debtors, financial accounts and items pending. Inside liabilities, on the other hand, we can consider capital and reserves, provisions, medium and long term debt, and current liabilities. At a second level we could consider within fixed assets, tangibles, intangibles, investments and deferred expenses. The list could be multiplied. These control groups are useful for generating reports of various types.

(II) Closing accounts

Another instance of a quotient system is when temporary accounts in an accounting system are closed. Consider a system $\mathcal{A} = (A|\ T|\ B)$ where the account set A is partitioned into equity accounts, permanent accounts and temporary accounts

$$A = A_e \cup A_p \cup A_t;$$

then the temporary accounts are divided into revenue and expense accounts, say $A_t = A_{tr} \cup A_{te}$. Thus there is a partition

$$A = A_e \cup A_p \cup A_{tr} \cup A_{te}.$$

Suppose that the temporary accounts are to be closed and their balances combined and added to the retained earnings account. There are two steps in the procedure to model this operation by quotient systems. Let E_1 be the equivalence relation on A corresponding to the partition above. In the quotient system $\mathcal{A}/E_1$ the temporary revenue and temporary expense accounts have now been combined into two accounts, while other accounts are unaffected.

The second step in the modeling process involves an equivalence relation E_2 on $\overline{A}_{E_1}$, with a corresponding partition of the accounts of $\mathcal{A}/E_1$ in which the two combined temporary accounts and retained earnings form one subset and thus are merged in the new system $(\mathcal{A}/E_1)/E_2$. The modeling procedure is summarized by the sequence

$$\mathcal{A} \to \mathcal{A}/E_1 \to (\mathcal{A}/E_1)/E_2 :$$

it is shown in 5.4 below that "doubledecker" quotient systems of this type may be identified with a single quotient system of $\mathcal{A}$.

It should be pointed out that the normal procedure of closing temporary accounts at the end of an accounting period is more complex than that just described since it may involve additional adjustments to account entries.

The hierarchy of quotient systems

The quotient systems of a given accounting system $\mathcal{A} = (A|\,T\,|\,B)$ correspond to equivalence relations on A, and so to unordered partitions of A. There is a well-established combinatorial theory of partitions of a set of distinct objects and this can be applied to the study of quotient systems of $\mathcal{A}$.

By use of the Inclusion-Exclusion Principle, it can be shown that the number of unordered partitions of a set with m distinct objects into k non-empty subsets is given by the formula

$$S(m,k) = \frac{1}{k!} \sum_{i=0}^{k} (-1)^i \binom{k}{i} (k-i)^m.$$

The number $S(m,k)$ is called a *Stirling number of the second kind*. It follows that the total number of partitions of a set of m distinct objects equals

$$B_m = \sum_{k=1}^{m} S(m,k),$$

which is known as the mth *Bell number*. (For an account of these results from combinatorics see [2]).

On applying the above facts to accounting systems, we conclude that the following holds.

(5.2.1). *The number of quotient systems of an accounting system with n accounts equals B_n. The number of such quotient systems with exactly k accounts is $S(n, k)$.*

As n increases, the total number of quotient systems increases rapidly. For example, when $n = 7$, the number of quotient systems is $B_7 = 877$. Of course, relatively few of these quotient systems will correspond to practical accounting procedures.

The partial ordering of quotient systems

There is a natural partial order on the set of equivalence relations on a given set, and hence on the set of quotient systems of an accounting system. Let E_1 and E_2 be equivalence relations on the account set A of an accounting system $\mathcal{A}$. This, of course, means that E_1 and E_2 are subsets of $A \times A$. We define an order $\leq$ on the set of equivalence relations by reverse set inclusion: thus

$$E_1 \leq E_2 \iff E_2 \subseteq E_1,$$

i.e., E_2-equivalence implies E_1-equivalence. Clearly $\leq$ is a partial order on the set of equivalence relations.

Now use this partial order to define an order on the set of quotients of $\mathcal{A}$, also written $\leq$,

$$\mathcal{A}/E_2 \leq \mathcal{A}/E_1 \iff E_1 \leq \cdot E_2 \iff E_2 \subseteq E_1.$$

Again this is a partial order, so that the quotient systems of $\mathcal{A}$ form a partially ordered set. Notice that the smaller the subset E of $A \times A$, the more accounts there are in the quotient $\mathcal{A}/E$. The largest possibility for the quotient $\mathcal{A}/E$ occurs when E is the set $\{(a, a) \mid a \in \mathcal{A}\}$, i.e., E is equality. In this case $\mathcal{A}/E$ is essentially the same system as $\mathcal{A}$. The smallest possibility for $\mathcal{A}/E$ is when $E = A \times A$, i.e., all accounts are equivalent, and $\mathcal{A}/E$ has a single account.

There are two natural binary operations called *meet* and the *join* which can be applied to equivalence relations on A. If E_1 and E_2

are equivalence relations on A, their *meet* is just the intersection

$$E_1 \wedge E_2 = E_1 \cap E_2,$$

which is also an equivalence relation on A. However, the union $E_1 \cup E_2$ is not in general an equivalence relation since the transitive law may fail to hold. To obtain an equivalence relation we must pass to the *transitive closure* of the relation $E_1 \cup E_2$. Thus we define the *join* of E_1 and E_2 to be

$$E_1 \vee E_2 = \text{the transitive closure of } E_1 \cup E_2.$$

Recall here that the *transitive closure* of a relation R is the smallest transitive relation containing R, which is just

$$R \cup R^2 \cup R^3 \cup \cdots.$$

Here R^k is defined recursively by the rule: $a\ R^k\ b$ if and only if there exists a c such that $a\ R^{k-1}c$ and $c\ R\ b$. It can be verified that meet and join play the roles of greatest lower bound and least upper bound in the partially ordering of equivalence relations on A. Consequently we have a *lattice*. The conclusion for accounting systems is:

(5.2.2). *The set of quotient systems of an accounting system $\mathcal{A}$ is a lattice with respect to the partial order $\leq$ where $\mathcal{A}/E_2 \leq \mathcal{A}/E_1$ if and only if $E_2 \subseteq E_1$.*

Example (5.2.3).

Let $\mathcal{A} = (A|\ T|\ B)$ be a system with four accounts a_1, a_2, a_3, a_4. Since $B_4 = 15$, there are exactly 15 quotient systems of $\mathcal{A}$, corresponding to the partitions of the account set A. For example, the partition

$$A = \{a_1, a_3\} \cup \{a_2\} \cup \{a_4\}$$

leads to a quotient system with three accounts.

5.3. Homomorphisms of Accounting Systems

It is a basic method in algebra to examine functions between two algebraic structures of the same type which relate the internal algebraic properties of the structures. Such functions are known as *homomorphisms*. As examples we mention homomorphisms of groups and modules, structures which have arisen in earlier chapters.

The importance of homomorphisms stems from the fact that they provide a means of comparing algebraic structures. Because of the widespread applicability of homomorphisms in abstract algebra, it is natural to attempt to construct a theory of homomorphisms between accounting systems. Such homomorphisms provide ways of comparing the accounting systems of different organizations, and can also represent certain commonly used operations on accounting systems.

The definition of a homomorphism

Consider two accounting systems

$$\mathcal{A} = (A| \ T| \ B) \text{ and } \mathcal{A}' = (A'| \ T'| \ B')$$

over the same ordered domain R, with account sets $A = \{a_1, a_2, \ldots, a_n\}$ and $A' = \{a'_1, a'_2, \ldots, a'_{n'}\}$. To define a homomorphism from $\mathcal{A}$ to $\mathcal{A}'$ we start with a function between the account sets

$$\theta : A \to A'.$$

This is used to generate a function between balance modules

$$\theta^* : \text{Bal}_n(R) \to \text{Bal}_{n'}(R)$$

where $\theta^*(\mathbf{v})$ is defined by the rule

$$(\theta^*(\mathbf{v}))_i = \begin{cases} \displaystyle\sum_{\theta(a_j)=a'_i} v_j \\ 0 \quad \text{if } a'_i \notin \text{Im}(\theta) \end{cases}$$

for $i = 1, 2, \ldots, n'$. Recall here that the *image* of the function θ, $\text{Im}(\theta)$, is the set $\{\theta(a_i)| \ i = 1, 2, \ldots, n\}$. In the definition the sum is to be formed over all j for which $\theta(a_j) = a'_i$. Thus the function θ^* sums all entries of $\mathbf{v}$ that correspond to accounts mapped by θ to the same account in A' and it assigns an entry of zero to any account of A' which is not in $\text{Im}(\theta)$. Notice that $\theta^*(\mathbf{v})$ is a balance vector.

It is a simple matter to deduce from the definition that

$$\theta^*(\mathbf{u} + \mathbf{v}) = \theta^*(\mathbf{u}) + \theta^*(\mathbf{v}) \quad \text{and} \quad \theta^*(r\mathbf{v}) = r\theta^*(\mathbf{v})$$

for all $\mathbf{u}, \mathbf{v} \in \mathrm{Bal}_n(R)$ and $r \in R$, equations which show that the function $\theta^* : \mathrm{Bal}_n(R) \to \mathrm{Bal}_{n'}(R)$ is a homomorphism of R-modules.

Before giving the formal definition of a homomorphism from $\mathcal{A}$ to $\mathcal{A}'$, we present an example illustrating the formation of the function θ^*.

Example (5.3.1).

Let $A = \{a_1, a_2, a_3, a_4, a_5\}$ and $A' = \{a'_1, a'_2, a'_3, a'_4\}$. A function $\theta : A \to A'$ is defined by the rules

$$\theta(a_1) = a'_1, \ \theta(a_2) = a'_3, \ \theta(a_3) = a'_2, \ \theta(a_4) = a'_2, \ \theta(a_5) = a'_2.$$

Then $\theta^* : \mathrm{Bal}_5(R) \to \mathrm{Bal}_4(R)$ sends

$$\begin{bmatrix} v_1 \\ v_2 \\ v_3 \\ v_4 \\ -v_1 - v_2 - v_3 - v_4 \end{bmatrix} \quad \text{to} \quad \begin{bmatrix} v_1 \\ -v_1 - v_2 \\ v_2 \\ 0 \end{bmatrix}.$$

Here account a'_4 gets the value 0 since $a'_4 \notin \mathrm{Im}(\theta) = \{a'_1, a'_2, a'_3\}$.

After this example we are ready to define a homomorphism from system $\mathcal{A} = (A|\ T|\ B)$ to system $\mathcal{A}' = (A'|\ T'|\ B')$. As before we start with a function $\theta : A \to A'$. Then θ is said to determine a *homomorphism of accounting systems*, written

$$\theta : \mathcal{A} \to \mathcal{A}',$$

provided that:

1. if $\mathbf{v} \in T$, then $\theta^*(\mathbf{v}) \in T'$, i.e., $\theta^*(T) \subseteq T'$;

2. if $\mathbf{b} \in B$, then $\theta^*(\mathbf{b})|_{\mathrm{Im}(\theta)} = \mathbf{b}'|_{\mathrm{Im}(\theta)}$ for some $\mathbf{b}' \in B'$.

What this means is that θ^* sends allowable transactions of $\mathcal{A}$ to allowable transactions of $\mathcal{A}'$ which affect only accounts in $\mathrm{Im}(\theta)$; on the other hand, θ^* sends an allowable balance vector of $\mathcal{A}$ to a balance vector that agrees in its $\mathrm{Im}(\theta)$-entries with some allowable balance vector of $\mathcal{A}'$. Notice that in condition (2) the vector $\mathbf{b}'$ might have non-zero entries for accounts in $A' \setminus \mathrm{Im}(\theta)$ and $\theta^*(\mathbf{b})$ might not belong to B'.

Monomorphisms, epimorphisms, isomorphisms

There are three special types of homomorphisms which are of importance. Suppose that $\theta : \mathcal{A} \to \mathcal{A}'$ is a homomorphism of accounting systems. If the function $\theta : A \to A'$ is injective, i.e., distinct accounts in A are sent to distinct accounts in A', then θ is called a *monomorphism*. This means that $|A| \leq |A'|$, so $\mathcal{A}'$ has at least as many accounts as $\mathcal{A}$. Another point to notice is that $(\theta^*(\mathbf{v}))_i = v_j$ if $\theta(a_j) = a_i'$ and this is zero if $a_i' \notin \mathrm{Im}(\theta)$. It follows that θ^* is injective and hence is a monomorphism of R-modules. Also an allowable transaction of $\mathcal{A}$ with zeros inserted for accounts in $A' \backslash \mathrm{Im}(\theta)$ becomes an allowable transaction of $\mathcal{A}'$. In addition each allowable balance vector of $\mathcal{A}$ is the restriction to $\mathrm{Im}(\theta)$ of some allowable balance vector of $\mathcal{A}'$. This type of homomorphism arises when new accounts are added to an accounting system – see Example 5.3.2 below.

The next special type of homomorphism $\theta : \mathcal{A} \to \mathcal{A}'$ is one for which the function $\theta : A \to A'$ is surjective. Under these circumstances each account in A' is the image under θ of an account in A and $|A| \geq |A'|$. It is easy to see that θ^* is surjective and hence it is an epimorphism of R-modules. If in addition

$$\theta^*(T) = T' \quad \text{and} \quad \theta^*(B) = B',$$

then $\theta : \mathcal{A} \to \mathcal{A}'$ is called an *epimorphism* of accounting systems. An important example of an epimorphism arises when a quotient of an accounting system is formed — see Example 5.3.3 below.

Finally, a homomorphism $\theta : \mathcal{A} \to \mathcal{A}'$ is called an *isomorphism* if it is both a monomorphism and an epimorphism. Under these circumstances $\theta : A \to A'$ is a bijection and it sets up a one-one correspondence between the accounts, allowable transactions and allowable balances of $\mathcal{A}$ and the corresponding entities of $\mathcal{A}'$. If there is at least one isomorphism from $\mathcal{A}$ to $\mathcal{A}'$, then $\mathcal{A}$ and $\mathcal{A}'$ is called *isomorphic* systems and the notation

$$\mathcal{A} \simeq \mathcal{A}'$$

is used. The essential observation is that isomorphic accounting systems are subject to the same rules of operation, although their account sets may be different. If $\theta : \mathcal{A} \to \mathcal{A}'$ is an isomorphism of accounting systems, it has an *inverse*, namely the homomorphism

$\theta^{-1} : \mathcal{A}' \to \mathcal{A}$ which arises from the set function $\theta^{-1} : A' \to A$, the inverse of the bijection θ.

Automorphisms

Following a widely used terminology in algebra, we call an isomorphism from an accounting system $\mathcal{A}$ to itself an *automorphism* of $\mathcal{A}$. The set of all automorphisms of $\mathcal{A}$ is denoted by

$$\mathrm{Aut}(\mathcal{A}).$$

Observe that an automorphism is a permutation of the account set A and thus $\mathrm{Aut}(\mathcal{A})$ is a subset of *the symmetric group* $\mathrm{Sym}(A)$, which consists of all permutations of A and whose group operation is functional composition. Now if α is an automorphism of $\mathcal{A}$, then by definition $\alpha \in \mathrm{Sym}(A)$ has the properties $\alpha^*(T) = T$ and $\alpha^*(B) = B$, so that the sets of allowable vectors of $\mathcal{A}$ are permuted by means of the action of the permutation α on vector entries. Conversely, any permutation α of A such that $\alpha^*(T) = T$ and $\alpha^*(B) = B$ gives rise to an automorphism of $\mathcal{A}$.

It is convenient to think of the automorphism α as permuting the integers $1, 2, \ldots, n$ in the same way as it permutes the accounts $a_1, a_2, \ldots, a_n$. With this convention the definition of α^* shows that $(\alpha^*(\mathbf{v}))_i = \mathbf{v}_{\alpha^{-1}(i)}$ since $\alpha^{-1}(a_i)$ is the unique account sent by α to a_i. If α, β are automorphisms, then

$$(\alpha \circ \beta)^* = \alpha^* \circ \beta^*.$$

For, if $\mathbf{v} \in \mathrm{Bal}_n(R)$, then

$$((\alpha \circ \beta)^*(\mathbf{v}))_i = \mathbf{v}_{(\alpha \circ \beta)^{-1}(i)} = \mathbf{v}_{\beta^{-1} \circ \alpha^{-1}(i)} = \mathbf{v}_{\beta^{-1}(\alpha^{-1}(i))},$$

while

$$(\alpha^* \circ \beta^*(\mathbf{v}))_i = (\beta^*(\mathbf{v}))_{\alpha^{-1}(i)} = \mathbf{v}_{\beta^{-1}(\alpha^{-1}(i))},$$

which establishes the claim.

From the equation just established it follows that the composite of two automorphisms is an automorphism. Now obviously the identity permutation on A is an automorphism, and the inverse of an automorphism is an automorphism since $\alpha^* \circ (\alpha^{-1})^*$ and $(\alpha^{-1})^* \circ \alpha$ equal the identity. Therefore we can assert that $\mathrm{Aut}(\mathcal{A})$ is a subgroup of the symmetric group $\mathrm{Sym}(A)$. The next result summarizes this discussion.

(5.3.1). *Let $\mathcal{A} = (A|\ T|\ B)$ be an accounting system. Then*

$$\mathrm{Aut}(\mathcal{A}) = \{\alpha \in \mathrm{Sym}(A) \mid \alpha^*(T) = T,\ \alpha^*(B) = B\}$$

and $\mathrm{Aut}(\mathcal{A})$ *is a subgroup of the symmetric group* $\mathrm{Sym}(A)$.

An automorphism group can be assigned to many algebraic structures, for example, groups, rings and modules. The importance of automorphism groups is that they tend to measure the amount of symmetry present in the structure. This is also the case with accounting systems. The more "symmetric" the sets of allowable vectors T and B, the larger will be the group $\mathrm{Aut}(\mathcal{A})$ where $\mathcal{A} = (A|\ T|\ B)$. For an extreme example suppose that $\mathcal{A}$ is the free accounting system, where $T = \mathrm{Bal}_n(R) = B$; then $\mathrm{Aut}(\mathcal{A})$ coincides with the whole symmetric group $\mathrm{Sym}(A)$.

One can envisage real-life situations where an accounting system has non-trivial automorphisms. For example, there might be two accounts that are subject to identical transaction and balance restrictions. Then interchanging the two accounts and fixing all others would lead to an automorphism of the accounting system. An example would be when two pieces of equipment are purchased at the same time for the same price and are subject to the same depreciation rules.

Examples of homomorphisms

We shall discuss in detail two sources of homomorphisms that will be present in most accounting systems.

Example (5.3.2). (*Monomorphisms and adjunction of accounts*)

A procedure that is standard for many accounting systems is the creation of new accounts. This gives rise to a monomorphism.

Let $\mathcal{A} = (A|\ T|\ B)$ be an accounting system with n accounts $a_1, a_2, \ldots, a_n$ and suppose that it is desired to create a number of new accounts $a_1^*, a_2^*, \ldots, a_m^*$. Then the new account set is

$$A' = \{a_1, a_2, \ldots, a_n, a_1^*, a_2^*, \ldots, a_m^*\},$$

with accounts in the order shown. Let

$$\theta : A \to A'$$

be the inclusion map, i.e. $\theta(a_i) = a_i$, for $i = 1, 2, \ldots, n$.

A new accounting system is to be created with account set A'; its allowable vectors should include $\theta^*(T)$ and $\theta^*(B)$. In practice one would expect there to be additional allowable transaction vectors that apply to the new accounts $a_1^*, a_2^*, \ldots, a_m^*$. Also it will be necessary to add further allowable balance vectors since those in $\theta^*(B)$ have zero entries corresponding to new accounts.

With these remarks in mind, we choose subsets T_1 and B_1 of $\mathrm{Bal}_{n+m}(R)$ which have zero entries for accounts $a_1, \ldots, a_n$ and put

$$T' = \theta^*(T) \cup T_1 \quad \text{and} \quad B' = \theta^*(B) \cup B_1.$$

Then the new accounting system is to be $\mathcal{A}' = (A' | \, T' | \, B')$. Thus $\mathcal{A}'$ has m additional accounts and it has additional allowable vectors, as well as those inherited from $\mathcal{A}$. Finally, the inclusion map $\theta : A \to A'$ gives rise to a homomorphism

$$\theta : \mathcal{A} \to \mathcal{A}'$$

since $\theta^*(T) \subseteq T'$ and $\theta^*(B) \subseteq B'$: clearly this is a monomorphism. Thus adjunction of the new accounts and allowable vectors to the existing accounting system sets up a monomorphism from the original system to the new system.

Example (5.3.3). (*Epimorphisms and combinations of accounts*)

An important way in which epimorphisms arise is when certain accounts in an accounting system are combined, i.e. a quotient system is formed. This is a situation familiar to every algebraist. From the accounting perspective it means that epimorphisms are capable of representing the procedures for generating reports on accounting systems.

Suppose that $\mathcal{A} = (A | \, T | \, B)$ is an accounting system and E is an equivalence relation on the account set A. Then, as was explained in 5.2, there is a corresponding quotient system

$$\mathcal{A}/E = (\overline{A} | \, \overline{T} | \, \overline{B})$$

where $\overline{A}$ is the set of E-equivalence classes and

$$\overline{T} = \{\overline{\mathbf{v}}_E \mid \mathbf{v} \in T\}, \quad \overline{B} = \{\overline{\mathbf{b}}_E \mid \mathbf{b} \in B\},$$

with $\overline{\mathbf{v}} = \overline{\mathbf{v}}_E$ being determined by the rule $(\overline{\mathbf{v}})_i = \overline{v}_i = \sum_{a_j E a_i} v_j$.

There is a natural surjective function

$$\sigma_E : A \to \overline{A}$$

defined by $\sigma_E(a_i) = [a_i]_E$, i.e., σ_E assigns to each account in A its E-equivalence class. Then, as was obsrved at the beginning on this section, σ_E induces a surjective R-module homomorphism

$$\sigma_E^* : \mathrm{Bal}_n(R) \to \mathrm{Bal}_{\overline{n}}(R)$$

where $\overline{n} = |\overline{A}| =$ the number of E-equivalence classes. The definition of σ_E^* shows that

$$(\sigma_E^*(\mathbf{v}))_i \;=\; \sum_{\sigma_E(a_j)\,=\,\sigma_E(a_i)} v_j \;=\; \sum_{a_j E a_i} v_j \;=\; \overline{v}_i,$$

and therefore $\sigma_E^*(\mathbf{v}) = \overline{\mathbf{v}}_E$. Consequently, $\sigma_E^*(T) = \overline{T}$ and $\sigma_E^*(B) = \overline{B}$, equations which show that

$$\sigma_E : \mathcal{A} \to \mathcal{A}/E$$

is an epimorphism of accounting systems. This is called the *canonical epimorphism* from $\mathcal{A}$ to $\mathcal{A}/E$: it is easy to remember its effect since this is to identify all accounts in the same E-equivalence class.

Take the simplest example, where E partitions the set A into asset, liability and equity accounts. The quotient system $\mathcal{A}/E$ has three accounts, total assets, total liabilities and total equity, and it represents the simplest type of report. It follows that there is an epimorphism from any accounting system with asset, liability and equity accounts to a 3-account system of this simple type.

The conclusions of the foregoing discussion are summarized in the following result.

(5.3.2). *Let $\mathcal{A} = (A|\,T|\,B)$ be an accounting system and let E be an equivalence relation on the account set A. Then the assignment $a \mapsto [a]_E$ determines an epimorphism $\sigma_E : \mathcal{A} \to \mathcal{A}/E$.*

5.4. Isomorphism Theorems

In a theory of homomorphisms between algebraic structures an algebraist would expect to find certain theorems called *isomorphism theorems* which relate homomorphisms with quotient structures. Such results are basic tools in many branches of algebra. Our purpose here is to formulate isomorphism theorems for accounting systems and to interpret these results from the point of view of accounting. The effect is to place the theory in a general algebraic setting.

The principal isomorphism theorem applies to a homomorphism of accounting systems $\theta : \mathcal{A} \to \mathcal{A}'$ and establishes an isomorphism between a certain quotient system of $\mathcal{A}$ and a subsystem of $\mathcal{A}'$ called the image of θ. In a sense this asserts that the essential information about any homomorphism from $\mathcal{A}$ is to be found within the quotient structures of $\mathcal{A}$.

Before proceeding to the statements of the theorems, it is necessary to examine two critical facets of a homomorphism, the associated equivalence relation and the image.

The equivalence relation associated with a homomorphism

Consider a homomorphism $\theta : \mathcal{A} \to \mathcal{A}'$ between accounting systems $\mathcal{A} = (A|\ T|\ B)$ and $\mathcal{A}' = (A'|\ T'|\ B')$ over an ordered domain R. This arises from a function between account sets $\theta : A \to A'$. An equivalence relation

$$E_\theta$$

on A is defined by the rule

$$a_i\, E_\theta\, a_j \iff \theta(a_i) = \theta(a_j).$$

Thus E_θ is associated with a partition of A in which accounts belonging to the same subset have the same image under θ. The equivalence relation E_θ can be used to construct the quotient system $\mathcal{A}/E_\theta$, which already suggests a close relationship between homomorphisms from $\mathcal{A}$ and quotients of $\mathcal{A}$.

The image of a homomorphism

Another important feature of a homomorphism of accounting systems $\theta : \mathcal{A} \to \mathcal{A}'$ is its *image*

$$\mathrm{Im}(\theta),$$

which is contained in $\mathcal{A}'$, although not necessarily as a subsystem. The account set of $\mathrm{Im}(\theta)$ is the image of A under the set function θ, also written $\mathrm{Im}(\theta)$ or $\theta(A) = \{\theta(a_i) \mid i = 1, 2, \ldots, n\}$. Let

$$\theta_0 : A \to \theta(A)$$

be the surjective function sending a_i to $\theta(a_i)$; thus θ_0 acts like θ, but it has smaller codomain. Then θ_0 induces a module homomorphism

$$\theta_0^* : \mathrm{Bal}_n(R) \to \mathrm{Bal}_{\overline{n}}(R)$$

where $n = |A|$, $\overline{n} = |\theta(A)|$ and θ_0^* is defined by the usual rule

$$(\theta_0^*(\mathbf{v}))_i = \sum_{\theta(a_j)=\theta(a_i)} v_j.$$

Here the sum is formed over all accounts with the same θ-value as a_i. Thus $\theta_0^*(T)$ and $\theta_0^*(B)$ are subsets of $\mathrm{Bal}_{\overline{n}}(R)$. The *image* of θ is defined to be the accounting system

$$\mathrm{Im}(\theta) = (\theta(A) \mid \theta_0^*(T) \mid \theta_0^*(B)).$$

Suppose that $\mathbf{v} \in T \cup B$; then $\theta_0^*(\mathbf{v})$ is a typical allowable vector for the accounting system $\mathrm{Im}(\theta)$. If $\mathbf{v} \in T$, then $\theta^*(\mathbf{v}) \in T'$ differs from $\theta_0^*(\mathbf{v})$ only through its $A'\backslash\theta(A)$-entries, all of which are zero. If $\mathbf{b} \in B$, then by definition of a homomorphism $\theta^*(\mathbf{b})|_{\mathrm{Im}(\theta)} = \mathbf{b}'|_{\mathrm{Im}(\theta)}$ where $\mathbf{b}' \in B'$. Again $\theta^*(\mathbf{b}) \in B'$ differs from $\theta_0^*(\mathbf{b})$ only in its $A'\backslash\theta(A)$-entries, but these need not be zero in this case.

From such considerations one might suspect that the $\mathrm{Im}(\theta)$ would be a subaccounting system of $\mathcal{A}'$: however, this is only true with additional conditions, reflecting the full force of the requirements for a subsystem given in 4.3.

(5.4.1). *Let $\theta : \mathcal{A} \to \mathcal{A}'$ be a homomorphism of accounting systems. Then $\mathrm{Im}(\theta)$, the image of θ, is a subsystem of $\mathcal{A}'$ if and only if the following conditions are satisfied:*

1. *If $\mathbf{v}' \in T' \cup B'$, then $\mathbf{v}'|_{\mathrm{Im}(\theta)}$ is a balance vector.*

2. *If $\mathbf{v}' \in T'$ and $\mathrm{sppt}(\mathbf{v}') \subseteq \mathrm{Im}(\theta)$, then $\mathbf{v}'|_{\mathrm{Im}(\theta)} \in \theta_0^*(T)$.*

3. *If $\mathbf{b}' \in B'$, then $\mathbf{b}'|_{\mathrm{Im}(\theta)} \in \theta_0^*(B)$.*

This is true because the conditions (1), (2), (3) are exactly what is needed to ensure that the image is a subsystem.

We are now ready to state the first of the isomorphism theorems.

(5.4.2). *Let $\theta : \mathcal{A} \to \mathcal{A}'$ be a homomorphism of accounting systems and let E_θ be the associated equivalence relation on the account set of $\mathcal{A}$. Then the assignment $[a]_{E_\theta} \mapsto \theta(a)$ induces an isomorphism $\psi : \mathcal{A}/E_\theta \to \mathrm{Im}(\theta)$.*

Proof
Write $\mathcal{A} = (A|\ T|\ B)$; then $\mathrm{Im}(\theta) = (\theta(A)|\ \theta_0^*(T)|\ \theta_0^*(B))$ where $\theta_0 : A \to \theta(A)$ is defined by $\theta_0(a_i) = \theta(a_i)$. Let $\mathcal{A}/E_\theta = (\overline{A}|\ \overline{T}|\ \overline{B})$ where $\overline{A}$ is the set of E_θ-equivalence classes $[a_i]_{E_\theta}$. A function between account sets

$$\psi : \overline{A} \to \theta(A)$$

is defined by $\psi([a_i]_{E_\theta}) = \theta(a_i)$. The first thing to observe is that $\theta(a_i)$ depends only on the equivalence class $[a_i]_{E_\theta}$, not on a_i, so ψ is a well-defined function. Next if $\psi([a_i]_{E_\theta}) = \psi([a_j]_{E_\theta})$, then $\theta(a_i) = \theta(a_j)$ and hence $[a_i]_{E_\theta} = [a_j]_{E_\theta}$. Therefore ψ is injective: since it is obviously surjective, ψ is a bijection.

It remains to show that ψ induces an homomorphism $\psi : \mathcal{A}/E_\theta \to \mathrm{Im}(\theta)$, for which purpose it suffices to prove that $\psi^*(\overline{T}) = \theta_0^*(T)$ and $\psi^*(\overline{B}) = \theta_0^*(B)$. Let $\overline{\mathbf{v}} \in \overline{T} \cup \overline{B}$; then by definition of the quotient system $\mathcal{A}/E_\theta$, there exists $\mathbf{v} \in T \cup B$ such that $\overline{\mathbf{v}} = \sigma_{E_\theta}^*(\mathbf{v})$, where $\sigma_{E_\theta} : \mathcal{A} \to \overline{A}$ is the canonical epimorphism defined in 5.3. Thus

$$\overline{v}_i = (\sigma_{E_\theta}^*(\mathbf{v}))_i = \sum_{\sigma_{E_\theta}(a_j) = \sigma_{E_\theta}(a_i)} v_j = \sum_{\theta_0(a_j) = \theta_0(a_i)} v_j = (\theta_0^*(\mathbf{v}))_i.$$

But

$$(\psi^*(\overline{\mathbf{v}}))_i = \sum_{\psi([a_j]) = \psi([a_i])} \overline{v}_j = \overline{v}_i$$

since ψ is injective. Therefore $(\psi^*(\overline{\mathbf{v}}))_i = (\theta_0^*(\mathbf{v}))_i$ for all i and thus

$$\psi^*(\overline{\mathbf{v}}) = \theta_0^*(\mathbf{v}).$$

Hence $\psi^*(\overline{T}) \subseteq \theta_0^*(T)$ and $\psi^*(\overline{B}) \subseteq \theta_0^*(B)$. Since the equation $\psi^*(\overline{\mathbf{v}}) = \theta_0^*(\mathbf{v})$ holds for any $\mathbf{v} \in T \cup B$ with $\overline{\mathbf{v}} = \sigma_{E_\theta}^*(\mathbf{v})$, we conclude that $\psi^*(\overline{T}) = \theta_0^*(T)$ and $\psi^*(\overline{B}) = \theta_0^*(B)$, which completes the proof. $\square$

There are two special cases of 5.4.2 which merit attention.

Example (5.4.1).

Suppose that $\theta : \mathcal{A} \to \mathcal{A}'$ is a monomorphism of accounting systems. Then $\overline{\mathcal{A}} = \mathcal{A}/E_\theta$ has account set $[a_i]_{E_\theta}$, $i = 1, 2, \ldots, n$, and $[a_i]_{E_\theta} = \{a_i\}$ since $\theta : A \to A'$ is injective. Thus the accounts of $\overline{\mathcal{A}}$ are the singleton sets $\{a_i\}$, $i = 1, 2, \ldots, n$. This means that $\mathcal{A}$ and $\overline{\mathcal{A}}$ are essentially the same system or, more precisely, they are isomorphic systems. According to 5.4.2, the accounting system $\overline{\mathcal{A}}$ is isomorphic with the image system $\text{Im}(\theta)$. Therefore $\mathcal{A}$ too is isomorphic with $\text{Im}(\theta)$ and so one can think of the monomorphism θ as "embedding" $\mathcal{A}$ in the larger system $\mathcal{A}'$. The procedure for adding accounts described in 5.3 is an instructive example of a monomorphism.

Example (5.4.2).

Let $\theta : \mathcal{A} \to \mathcal{A}'$ be an epimorphism of accounting system. By definition $\theta(A) = A'$, $\theta^*(T) = T'$ and $\theta^*(B) = B'$, so that $\text{Im}(\theta) = \mathcal{A}'$. Then 5.4.2 shows that $\mathcal{A}'$ is isomorphic with the quotient system $\mathcal{A}/E_\theta$.

Another indication of the importance of monomorphisms and epimorphisms is provided by the next result.

(5.4.3). *Every homomorphism between accounting systems may be expressed as the composite of an epimorphism followed by a monomorphism.*

Proof

Let $\theta : \mathcal{A} \to \mathcal{A}'$ be a homomorphism between accounting systems. Following the previous convention, we write E_θ for the equivalence relation on A determined by $\theta : A \to A'$ and σ_{E_θ} for the canonical homomorphism from $\mathcal{A}$ to $\mathcal{A}/E_\theta$ defined by the rule $\sigma_{E_\theta}(a_i) = [a_i]_{E_\theta}$: furthermore $\psi : \mathcal{A}/E_\theta \to \text{Im}(\theta)$ is the isomorphism in 5.4.2, so that $\psi([a_i]_{E_\theta}) = \theta(a_i)$. Let $\iota : \theta(A) \to A'$ denote the inclusion map; this induces a monomorphism of accounting systems $\iota : \text{Im}(\theta) \to \mathcal{A}'$ and it is clear that the composite

$$\iota \circ \psi : \mathcal{A}/E_\theta \to \mathcal{A}'$$

is a also monomorphism.

Next for any account a_j we have

$$(\iota \circ \psi) \circ \sigma_{E_\theta}(a_j) = \iota(\psi(\sigma_{E_\theta}(a_j))) = \iota(\psi([a_j]_{E_\theta})) = \iota(\theta(a_j)) = \theta(a_j).$$

Hence $\theta = (\iota \circ \psi) \circ \sigma_{E_\theta}$ and, since σ_{E_θ} is an epimorphism, the result is proved. $\square$

We remark that care must be exercised in forming composites of homomorphisms of accounting systems. Examples show that the composite of a monomorphism followed by an epimorphism need not be a homomorphism, which is in contrast to the statement of 5.4.3: the reason for this is that an allowable balance vector need not be mapped to the restriction of an allowable balance vector by the composite, as is required by the definition.

Quotients of quotient systems

The final isomorphism theorem gives information about quotient systems of quotient systems. It reveals that these potentially complex objects are in fact no worse than quotients of the original system. The accounting interpretation here is about reports on reports. In the accounting system of a large organization accounts may be grouped into control groups, and the control groups may themselves be grouped into control groups. The process of forming reports on the groups and on the subsidiary groups gives rise to the quotient of a quotient situation.

Let $\mathcal{A} = (A|\ T|\ B)$ be an accounting system and let E be an equivalence relation on the account set A. Then $\mathcal{A}/E = (\overline{A}|\ \overline{T}|\ \overline{B})$ has as its account set $\overline{A}$, the set of all E-equivalence classes $[a_i]_E$. Now suppose that F is an equivalence relation on $\overline{A}$, so that it is possible to form the quotient of the quotient system $\mathcal{A}/E$ by F, i.e.,

$$(\mathcal{A}/E)/F.$$

The key to understanding this complex object is the observation that E and F determine a new equivalence relation on A denoted by

$$E \mathbin{\#} F,$$

where by definition

$$a_i\,(E \mathbin{\#} F)\,a_j \iff [a_i]_E\,F\,[a_j]_E.$$

In terms of partitions, $E \mathbin{\#} F$ arises by taking the partition of A determined by E and forming the union of all subsets in this partition that belong to same subset in the partition corresponding to F. This procedure leads to a partition with larger subsets than E

which determines the equivalence relation $E \# F$. We can therefore form the quotient system

$$\mathcal{A}/(E \# F).$$

The connection with quotients of quotients is shown by:

(5.4.4). *Let $\mathcal{A} = (A| \, T| \, B)$ be an accounting system. Suppose that $\mathcal{A}/E$ is a quotient of $\mathcal{A}$ and $(\mathcal{A}/E)/F$ a quotient of $\mathcal{A}/E$. Then*

$$(\mathcal{A}/E)/F \simeq \mathcal{A}/(E \# F).$$

Proof
Write $\mathcal{A}/E = (\overline{A}| \, \overline{T}| \, \overline{B})$ and $\mathcal{A}/(E \# F) = (\overline{\overline{A}}, \overline{\overline{T}}, \overline{\overline{B}})$. We begin the proof by introducing a function $\theta : \overline{A} \to \overline{\overline{A}}$, which is defined by the rule

$$\theta([a_i]_E) = [a_i]_{E \# F}.$$

This is in fact a well-defined function since $a_j \, E \, a_i$ implies that $[a_i]_E \, F \, [a_j]_E$, i.e., $[a_i]_{E \# F} = [a_j]_{E \# F}$. It is clear that θ is surjective; we will show that θ induces an epimorphism from $\mathcal{A}/E$ to $\mathcal{A}/(E \# F)$. Let $\mathbf{v} \in \mathrm{Bal}_n(R)$ where $n = |A|$ and write $\overline{\mathbf{v}} = \sigma_E^*(\mathbf{v})$, so that $\overline{v}_j = \sum_{a_k E a_j} v_k$. Since $\theta([a_i]_E) = [a_i]_{E \# F}$, we have

$$(\theta^*(\overline{\mathbf{v}}))_i = \sum_{a_j (E \# F) a_i} \overline{v}_j = \sum_{[a_j]_E \, F \, [a_i]_E} \Big(\sum_{a_k E a_j} v_k \Big).$$

In the right hand sum the entries v_k are first summed over the E-equivalence class of a_j, and then these sums are added up over the F-equivalence class of $[a_i]_E$. Therefore, by definition of $E \# F$,

$$\sum_{[a_j]_E \, F \, [a_i]_E} \Big(\sum_{a_k E a_j} v_k \Big) = \sum_{a_k (E \# F) a_i} v_k = (\sigma_{E \# F}^*(\mathbf{v}))_i$$

for all i. It follows that $\theta^*(\overline{\mathbf{v}}) = \sigma_{E \# F}^*(\mathbf{v})$ and hence that $\theta^* \circ \sigma_E^* = \sigma_{E \# F}^*$ because $\overline{\mathbf{v}} = \sigma_E^*(\mathbf{v})$. Next we have $\sigma_E^*(T) = \overline{T}$ and $\sigma_E^*(B) = \overline{B}$. In addition $\sigma_{E \# F}^*(T) = \overline{\overline{T}}$ and $\sigma_{E \# F}^*(\overline{B}) = \overline{\overline{B}}$, so we can deduce that

$$\theta^*(\overline{T}) = (\theta^* \circ \sigma_E^*)(T) = \sigma_{E \# F}^*(T) = \overline{\overline{T}},$$

and in a similar way $\theta^*(\overline{B}) = \overline{\overline{B}}$. Hence θ induces an epimorphism of accounting systems $\theta : \mathcal{A}/E \to \mathcal{A}/(E\#F)$. By 5.4.2 we obtain $(\mathcal{A}/E)/E_\theta \simeq \mathcal{A}/(E\#F)$.

In order to complete the proof we need only show that $E_\theta = F$. Now $[a_i]_E \; E_\theta \; [a_j]_E$ holds precisely when $\theta([a_i]_E) = \theta([a_j]_E)$, i.e., $[a_i]_{E\#F} = [a_j]_{E\#F}$ by definition of θ; this is equivalent to $[a_i]_E \; F \; [a_j]_E$, from which we deduce that $E_\theta = F$. $\square$

Since the last proof is rather complicated, we will illustrate it with an example.

Example (5.4.3).

Consider an accounting system $\mathcal{A} = (A|\; T|\; B)$ over $\mathbb{Z}$ with six accounts a_i, $i = 1, 2, \ldots, 6$. Let E be the equivalence relation on A with partition

$$\{a_1, a_2, a_3\}, \; \{a_4, a_5\}, \; \{a_6\}.$$

The quotient system $\mathcal{A}/E$ has three accounts

$$\overline{a}_1 = \{a_1, a_2, a_3\}, \; \overline{a}_2 = \{a_4, a_5\}, \; \overline{a}_3 = \{a_6\}.$$

Let F be the equivalence relation on the account set of $\mathcal{A}/E$ with partition

$$\{\overline{a}_1, \overline{a}_3\}, \; \{\overline{a}_2\}.$$

Then $(\mathcal{A}/E)/F$ has two accounts

$$a_1' = \{\overline{a}_1, \overline{a}_3\}, \; a_4' = \{\overline{a}_2\}.$$

Notice that the accounts of $(\mathcal{A}/E)/F$ are sets of sets of accounts of $\mathcal{A}$, so this system is a complex object. However, $(\mathcal{A}/E)/F$ is isomorphic with the system $\mathcal{A}/(E\#F)$ by 5.4.4. Now the equivalence relation $E\#F$ on A has partition

$$\{a_1, a_2, a_3, a_6\}, \; \{a_4, a_5\}$$

so, as expected, $\mathcal{A}/(E\#F)$ also has two accounts

$$\overline{\overline{a}}_1 = \{a_1, a_2, a_3, a_6\}, \; \overline{\overline{a}}_4 = \{a_4, a_5\}.$$

Next let $\mathbf{v} \in \mathrm{Bal}_6(\mathbb{Z})$ have entries

$$v_1, \; v_2, \; v_3, \; v_4, \; v_5, \; v_6 = -\sum_{i=1}^{5} v_i.$$

Now we pass to successive quotients $\mathcal{A}/E$ and $(\mathcal{A}/E)/F$, computing the images of $\mathbf{v}$. Thus $\sigma_E^*(\mathbf{v})$ has entries $v_1 + v_2 + v_3$, $v_4 + v_5$, v_6, and $\sigma_F^*(\sigma_E^*(\mathbf{v}))$ has entries $-v_4 - v_5$, $v_4 + v_5$. Finally, observe that $\sigma_{E\#F}^*(\mathbf{v})$ has the same entries as $\sigma_F^*(\sigma_E^*(\mathbf{v}))$, as we expect from 5.4.4.

In conclusion we remark that this chapter is more abstract than most others in the book. However, it should be stressed that the isomorphism theorems which have been established have interpretations that give insight into the operation of accounting systems. They lend credence to the role of algebra in accounting theory and serve to affirm the position of the algebraic theory of accounting within the general area of applied algebra. A final note: algebraists may have noticed the absence of one type of isomorphism theorem that is found in many parts of algebra. This is a result asserting that the image of a subsystem is isomorphic with a subsystem of the image. In general there is no such result in accounting theory since the image of a homomorphism need not be a subsystem.

Chapter Six

Accounting Systems and Automata

An automaton is a theoretical device which can simulate the operation of a digital computer. The algebraic theory of automata had its origins in the researches of A.M. Turing and C. Shannon. Recently automata theory has been applied in such diverse fields as biology, psychology, biochemistry and sociology. It has also been applied to economics through systems theory (Ames [1983]) and, more recently, to finance (Cruz Rambaud and García Pérez [2001]). It turns out that the accounting process can be described in terms of certain automata in which the state of an accounting system is transformed through the action of inputs, i.e., transactions, giving rise to specific outputs providing information about the system. The aim of this chapter is to lay out an approach to accounting using the concept of an automaton and to show how the operation of the double-entry bookkeeping system can be modeled by using this mathematical concept. The chapter begins with a brief introduction to automata theory.

6.1. Introduction to Semiautomata and Automata

A *semiautomaton* is a triple

$$\mathcal{S} = (Z, X, \delta),$$

consisting of two non-empty sets Z and X and a function

$$\delta : Z \times X \longrightarrow Z.$$

Here the set Z is called the *set of states*, X the *input alphabet* and δ the *next state function* of $\mathcal{S}$. The semiautomaton functions in the following manner: if the semiautomaton is in a state $z \in Z$ and it reads an input symbol $x \in X$, then it moves to a new state $\delta(z, x) \in Z$.

A more complex concept is that of an *automaton*, by which is meant a quintuple

$$\mathcal{M} = (Z, X, Y, \delta, \lambda)$$

where (Z, X, δ) is a semiautomaton, Y is a non-empty set called the *output alphabet* and

$$\lambda : Z \times X \longrightarrow Y$$

is a function called the *output function*. The automaton operates as follows. If the automaton is in state $z \in Z$ and it reads an input symbol $x \in X$, then moves to the new state $\delta(z, x)$ and it prints the output symbol $\lambda(z, x)$. Observe that a semiautomaton can be regarded as an automaton in which the states serve as the output symbols.

It is helpful to think of the automaton as a box with a head which can read symbols on an input tape and print symbols on an output tape. At any instant the automaton is in some state; after reading an input symbol, it prints a symbol on the output tape and moves to another state.

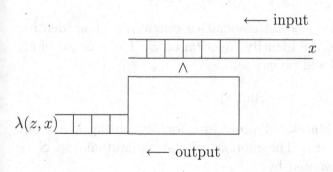

The digraph of an automaton

Automata and semiautomata can be represented by labeled digraphs. Take the case of a semiautomaton with state set Z, input alphabet X and next state function δ: the states are represented by vertices of the digraph and there is a directed edge from z to z' with a label x if $z' = \delta(z, x)$. If the semiautomaton is an automaton, the output $\lambda(z, x)$ can be represented by an additional arrow drawn from the initial state.

- The case of a semiautomaton:

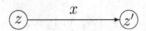

- The case of an automaton:

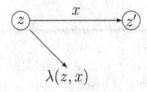

The monoid of a semiautomaton

A *monoid* is an algebraic structure consisting of a set equipped with an associative binary operation and an identity element. A standard example of a monoid is the set of all functions $\sigma : S \longrightarrow S$ on a non-empty set S where the binary operation is functional composition

$$\sigma_1 \circ \sigma_2(s) = \sigma_1(\sigma_2(s)) \, , s \in S,$$

which is well-known to be an associative operation. The identity element is, of course, the identity function on S. The monoid of all functions on a set S will be written

$$\text{Fun}(S).$$

There is a well established procedure for associating a monoid with a semiautomaton. The monoid of the semiautomaton $\mathcal{S} = (Z, X, \delta)$, which is denoted by

$$\text{Mon}(\mathcal{S}),$$

is defined as follows. If $x \in X$, there is a function $f_x : Z \to Z$ which is defined by using the next state function of $\mathcal{S}$,

$$f_x(z) = \delta(z, x), \quad z \in Z.$$

Thus $f_x(z)$ is the next state of the semiautomaton if it is initially in state z and it reads the input symbol x. Of course $f_x \in \mathrm{Fun}(Z)$. Now define the *monoid of the semiautomaton $\mathcal{S}$* to be the *submonoid* of $\mathrm{Fun}(Z)$ generated by all the functions f_x where $x \in X$:

$$\mathrm{Mon}(\mathcal{S}) = \mathrm{Mon}\langle f_x |\ x \in X \rangle.$$

This means that each element of $\mathrm{Mon}(\mathcal{S})$ is the composite of a finite sequence of functions,

$$f_{x_1} \circ f_{x_2} \circ \cdots \circ f_{x_k}, \quad x_i \in X.$$

Then $\mathrm{Mon}(\mathcal{S}) \subseteq \mathrm{Fun}(Z)$ and $\mathrm{Mon}(\mathcal{S})$ is a submonoid of the monoid $\mathrm{Fun}(Z)$.

Extension to a free monoid

It is common practice to regard a sequence of inputs $x_1, x_2, \ldots, x_r$ as acting on the states of a semiautomaton or automaton. A concrete example is an elevator: here the floors of a building are the states and the elevator can receive and act on a sequence of messages, not just a single one. This concentrates our attention on the set of sequences of elements from the input alphabet X. In algebra there is a tool which allows us to describe this type of object, namely the free monoid on X.

Define

$$\overline{X}$$

to be the set of all *words*, i.e., finite sequences of elements in X, written in the form

$$x_1 x_2 \ldots x_r :$$

the *empty word* Λ is the case $r = 0$. A binary operation $*$ on $\overline{X}$ is defined by concatenation. This means that if $\overline{x} = x_1 x_2 \ldots x_p$ and $\overline{x}' = x_1' x_2' \ldots x_q'$ are two elements of $\overline{X}$ of lengths p and q respectively: then

$$\overline{x} * \overline{x}' = x_1 x_2 \ldots x_p x_1' x_2' \ldots x_q'.$$

Thus the operation $*$ adjoins $\overline{x}'$ to $\overline{x}$ on the right, producing a sequence of length $p + q$. Obviously $*$ is an associative operation on $\overline{X}$ and by convention the empty word Λ is the identity element, i.e.,

$$\overline{x} * \Lambda = \overline{x} = \Lambda * \overline{x}$$

for all $\overline{x} \in \overline{X}$. Then $\overline{X}$ is a monoid called the *free monoid on X*.

Returning to our study of the semiautomaton $\mathcal{S} = (Z, X, \delta)$, we extend the input set X to the free monoid $\overline{X}$ on it, with Λ as the identity element. We can also extend the next state function δ to a function $\overline{\delta} : Z \times \overline{X} \to Z$ by the following recursive definition. Given $z \in Z$ and $x_1, x_2, \ldots, x_r \in X$, the state $\overline{\delta}(z, x_1 x_2 \ldots x_r)$ is computed by the rules

$$\overline{\delta}(z, \Lambda) = z,$$

$$\overline{\delta}(z, x_1) = \delta(z, x_1),$$

$$\overline{\delta}(z, x_1 x_2 \ldots x_r) = \overline{\delta}(\delta(z, x_1), x_2 \ldots x_r),$$

where $r > 1$. Thus once the action of the sequence of length $r - 1$ on the states has been defined, the action of a sequence of length r is determined by the preceding equations. In this way the semiautomaton $\mathcal{S} = (Z, X, \delta)$ has been extended to a semiautomaton

$$\overline{\mathcal{S}} = (Z, \overline{X}, \overline{\delta}).$$

In the case of an automaton $\mathcal{M} = (Z, X, Y, \delta, \lambda)$, the action of an input sequence $x_1 x_2 \ldots x_r$ on an initial state z produces as an output a sequence $y_1 y_2 \ldots y_r$ of elements of the output alphabet Y which is determined by a new output function $\overline{\lambda} : Z \times \overline{X} \to \overline{Y}$ in a similar fashion:

$$\overline{\lambda}(z, \Lambda) = \Lambda,$$

$$\overline{\lambda}(z, x_1) = \lambda(z, x_1),$$

$$\overline{\lambda}(z, x_1 x_2 \ldots x_r) = \lambda(z, x_1)\overline{\lambda}(\delta(z, x_1), x_2 \ldots x_r),$$

where $r > 1$. Notice that in all these definitions the automaton reads the input symbols in the order $x_1, x_2, \ldots, x_r$. The automaton $\mathcal{M} = (Z, X, Y, \delta, \lambda)$ has therefore been extended to the new automaton

$$\overline{\mathcal{M}} = (Z, \overline{X}, \overline{Y}, \overline{\delta}, \overline{\lambda}).$$

The operation of the semiautomaton $\overline{\mathcal{S}} = (Z, \overline{X}, \overline{\delta})$ can be visualised in the following way. Let $z_0 \in Z$ be the initial state and let $x_1 x_2 \ldots x_r \in \overline{X}$ be the input sequence. The semiautomaton passes to successive states $z_1, z_2, \ldots, z_r$ where $z_i = \delta(z_{i-1}, x_i)$, $i \geq 1$, the final state being z_r.

In the case of the automaton $\overline{\mathcal{M}} = (Z, \overline{X}, \overline{Y}, \overline{\delta}, \overline{\lambda})$ initially in state z_0, when the input sequence $x_1 x_2 \ldots x_r$ is read, the output sequence is $y_1 y_2 \ldots y_r$ where $y_i = \lambda(z_{i-1}, x_i)$, $i \geq 1$. Thus the automaton can be represented by the diagram below.

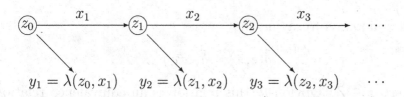

Next we consider the connection between the monoid of a semiautomaton $\mathcal{S} = (Z, X, \delta)$ and that of its extension $\overline{\mathcal{S}} = (Z, \overline{X}, \overline{\delta})$.

(6.1.1). *The monoids of a semiautomaton $\mathcal{S}$ and its extension $\overline{\mathcal{S}}$ are identical.*

Proof
Recall that $\mathrm{Mon}(\mathcal{S})$ is generated by all functions $f_x : Z \to Z$ where $x \in X$ and $f_x(z) = \delta(z, x)$. Similarly $\mathrm{Mon}(\overline{\mathcal{S}})$ is generated by all functions $\overline{f}_x : Z \to Z$ where $\overline{f}_x(z) = \overline{\delta}(z, x)$. From this one can prove that the following rule holds:

$$\overline{f}_{x_1 \ldots x_{r-1} x_r} = f_{x_r} \circ f_{x_{r-1}} \circ \cdots \circ f_{x_1},$$

where $x_i \in X$. Indeed by definition of the function $\overline{f}$, we have for any $z \in Z$ and $r \geq 2$

$$\overline{f}_{x_1 x_2 \ldots x_r}(z) = \overline{\delta}(z, x_1 x_2 \ldots x_r) = \overline{\delta}(\delta(z, x_1), x_2 \ldots x_r),$$

which equals

$$\overline{f}_{x_2 \ldots x_r}(\delta(z, x_1)) = \overline{f}_{x_2 \ldots x_r}(f_{x_1}(z)).$$

Therefore $\overline{f}_{x_1 x_2 \ldots x_r} = \overline{f}_{x_2 \ldots x_r} \circ f_{x_1}$ and the claim follows by induction on r. The result is now evident. $\qquad\square$

Equivalence relations associated with an automaton

There are two notable equivalence relations on the input set and the state set that can be applied to an automaton.

1. *An equivalence relation on inputs*

Consider an extended semiautomaton $\overline{\mathcal{S}} = (Z, \overline{X}, \overline{\delta})$, where the bar has the usual interpretation. We define a binary relation on $\overline{X}$ by saying that $\overline{x}$ and $\overline{x}' \in \overline{X}$ are equivalent if

$$\overline{f}_{\overline{x}} = \overline{f}_{\overline{x}'},$$

which is amounts to saying that

$$\overline{\delta}(z, \overline{x}) = \overline{\delta}(z, \overline{x}')$$

for every $z \in Z$. Obviously this relation is an equivalence relation on $\overline{X}$. In words two sequences of inputs are equivalent if, starting from the same state, they always produce the same new state.

2. *An equivalence relation on states*

Let $\overline{\mathcal{M}} = (Z, \overline{X}, \overline{Y}, \overline{\delta}, \overline{\lambda})$ be an extended automaton. Two states z, z' will be called equivalent if for every $\overline{x} \in \overline{X}$

$$\overline{\lambda}(z, \overline{x}) = \overline{\lambda}(z', \overline{x}).$$

Again it is clear that this an equivalence relation on Z. Thus two states are equivalent if they lead to the same output when the same input is read.

This concludes our introduction to automata theory: for a detailed account see [5]. In the following sections we will describe two ways in which the operation of an accounting system can be represented by an automaton. Both automata keep track of the transactions applied and the evolving balance sheet of the accounting system, while the second more complex automaton also controls the times at which a transaction can be performed.

6.2. Accounting Systems as Automata I

Recall from Chapter 4 that an accounting system over an ordered domain R is a triple

$$\mathcal{A} = (A \mid T \mid B)$$

where A is the set of accounts, T is the set of allowable transactions and B is a set of allowable balance vectors for the system. There is a straightforward way to represent the mode of operation of the accounting system by an automaton

$$\mathcal{M}_A$$

defined in the following way.

- The state set is B.

- The input set is $\mathrm{Bal}_n(R)$ where n is the number of accounts.

- The output set is $B \cup \{e_T, e_B\}$ where the symbols e_T and e_B stands for error messages, specifically

  - e_T: "the transaction is not allowable";
  - e_B: "the balance vector is not allowable".

- The next state function $\delta : B \times \mathrm{Bal}_n(R) \longrightarrow B$ is given by the rule:

$$\delta(\mathbf{b}, \mathbf{v}) = \begin{cases} \mathbf{b} + \mathbf{v} & \text{if } \mathbf{v} \in T \text{ and } \mathbf{b} + \mathbf{v} \in B \\ \mathbf{b} & \text{otherwise} \end{cases}$$

- The output function $\lambda : B \times \mathrm{Bal}_n(R) \longrightarrow B \cup \{e_T, e_B\}$ is given by the rule:

$$\lambda(\mathbf{b}, \mathbf{v}) = \begin{cases} \mathbf{b} + \mathbf{v} & \text{if } \mathbf{v} \in T \text{ and } \mathbf{b} + \mathbf{v} \in B \\ e_T & \text{if } \mathbf{v} \notin T \\ e_B & \text{if } \mathbf{v} \in T \text{ and } \mathbf{b} + \mathbf{v} \notin B \end{cases}$$

The automaton $\mathcal{M}_A$ functions in the following manner. Assume that the accounting system has balance vector $\mathbf{b}$ at some instant, so $\mathbf{b} \in B$. A transaction $\tau_\mathbf{v}$ is applied to the system where $\mathbf{v} \in \mathrm{Bal}_n(R)$. Suppose that $\mathbf{v} \in T$, i.e., the transaction is allowable; then the balance vector $\tau_\mathbf{v}(\mathbf{b}) = \mathbf{b} + \mathbf{v}$ is computed. If $\mathbf{b} + \mathbf{v}$ is in B, the new balance $\mathbf{b} + \mathbf{v}$ is allowable and it becomes the next state. The new balance is then printed on the output tape. If $\mathbf{v} \notin T$, the transaction is not allowable and is rejected: the state remains $\mathbf{b}$ and the error message e_T is printed on the output tape. Finally, if $\mathbf{v} \in T$ and $\mathbf{b} + \mathbf{v} \notin B$, the transaction is rejected since it would lead to a non-allowable balance. The automaton remains in state $\mathbf{b}$ and the error message e_B is printed on the output tape.

Remark.
As was observed in 4.2, it is reasonable to require that the set of allowable transactions contain the zero transaction: otherwise, when the zero vector is read by the automaton, an error message will be generated, even although the transaction does not change the system.

What the automaton $\mathcal{M}_\mathcal{A}$ achieves is the generation of a matrix which displays the financial history of the accounting system $\mathcal{A}$, the rows of the matrix corresponding to the accounts and the columns to the transactions: this is the balance matrix, which was introduced in 3.1. The first column of the matrix is the balance vector whose entries are the initial account balances. As each transaction is applied, a new column is generated, which is the current balance vector. The last column of the matrix gives the final account balances. As was mentioned in 3.1, the applied transactions can be recovered from the balance matrix sine the ith transaction applied to the system is obtained by subtracting the ith column from the $(i+1)$th; if the resulting difference is zero, then either a non-allowable transaction was applied or else a non-allowable balance vector was produced.

The automaton $\mathcal{M}_\mathcal{A}$ has the advantage of being quite simple and of displaying most of the data that one would normally require. In 6.3 we will introduce a second automaton associated with an accounting system which introduces the element of time, the so-called *time enhanced automaton* of the system.

The monoid of an accounting system

Let $\mathcal{A} = (A|\, T|\, B)$ be an accounting system; then the associated automaton $\mathcal{M}_\mathcal{A}$ has a monoid $\mathrm{Mon}(\mathcal{M}_\mathcal{A})$, which will be called the *monoid of the accounting system* $\mathcal{A}$ and denoted by

$$\mathrm{Mon}(\mathcal{A}).$$

By definition $\mathrm{Mon}(\mathcal{A})$ is generated by all functions of the form $f_\mathbf{v} : B \to B$, where $\mathbf{v} \in \mathrm{Bal}_n(R)$: recall that

$$f_\mathbf{v}(\mathbf{b}) = \delta(\mathbf{b}, \mathbf{v}), \quad (\mathbf{b} \in B).$$

Thus $f_\mathbf{v}(\mathbf{b}) = \mathbf{b} + \mathbf{v}$ provided that $\mathbf{v} \in T$ and $\mathbf{b} + \mathbf{v} \in B$, and otherwise $f_\mathbf{v}(\mathbf{b}) = \mathbf{b}$. Of course, $f_\mathbf{v} \in \mathrm{Fun}(B)$ and $\mathrm{Mon}(\mathcal{A})$ is a submonoid of the monoid $\mathrm{Fun}(B)$.

Each element of $\mathrm{Mon}(\mathcal{A})$ is the composite of a finite sequence of functions and has the form $f_{\mathbf{v}_1} \circ f_{\mathbf{v}_2} \circ \cdots \circ f_{\mathbf{v}_k}$. Note that if $\mathbf{v}_i$ is not allowable, then $f_{\mathbf{v}_i}$ is the identity function and thus can be deleted from the composite: consequently we can assume that each $\mathbf{v}_i$ in the composite belongs to T. In addition, $\mathbf{v}_i$ must yield an allowable balance vector, otherwise it will produce no effect on the balance vector. This discussion shows that in selecting generators for $\mathrm{Mon}(\mathcal{A})$ we can restrict ourselves to vectors $\mathbf{v}$ in T such that $\mathbf{b} + \mathbf{v}$ belongs to the set B for some $\mathbf{b} \in B$. Hence we can assume that $\mathbf{v}$ has the form $\mathbf{v} = \mathbf{c} - \mathbf{b}$ where $\mathbf{b}$ and $\mathbf{c}$ belong to B. This conclusion may be stated in the following form.

(6.2.1). *Let $\mathcal{A} = (A \mid T \mid B)$ be an accounting system. Then*

$$\mathrm{Mon}(\mathcal{A}) = \mathrm{Mon}\langle f_{\mathbf{v}} \mid \mathbf{v} \in (B - B) \cap T \rangle,$$

where $B - B$ denotes the set of all differences $\mathbf{c} - \mathbf{b}$ with $\mathbf{c}, \mathbf{b} \in B$. Hence every element of $\mathrm{Mon}(\mathcal{A})$ has the form $f_{\mathbf{v}_1} \circ f_{\mathbf{v}_2} \circ \cdots \circ f_{\mathbf{v}_k}$ where $\mathbf{v}_i \in (B - B) \cap T$.

Monoids of unbounded systems

With a general accounting system it can be difficult to understand the structure of its monoid, particularly when complex balance restrictions are present. It is worthwhile looking at the monoids of accounting systems in which the sets of allowable transactions and balances are large. While it might be objected that such systems are unrealistic, they do provide insight into the algebraic structure of the monoid and how its properties are related to those of the accounting system.

Example (6.2.1).
Consider an accounting system with no balance restrictions

$$\mathcal{A} = (A \mid T \mid \mathrm{Bal}_n(R)).$$

In this system all balance vectors are allowable: recall that such a system is called unbounded. If $\mathcal{A}$ is unbounded, then every transaction $\tau_{\mathbf{v}}$ with $\mathbf{v} \in T$ is accepted by the automaton $\mathcal{S}_{\mathcal{A}}$. It is convenient to write

$$\mathbf{v}'$$

for the function $\tau_{\mathbf{v}} : \mathrm{Bal}_n(R) \to \mathrm{Bal}_n(R)$; thus

$$\mathbf{v}'(\mathbf{b}) = \mathbf{b} + \mathbf{v}$$

for all $\mathbf{v}$, $\mathbf{b} \in \mathrm{Bal}_n(R)$. Notice that $\mathbf{v}' = f_{\mathbf{v}}$ if and only if $\mathbf{v} \in T$: however, these functions have different values at $\mathbf{v}$ if $\mathbf{v} \notin T$ since then $f_{\mathbf{v}}$ is the identity function.

If $\mathbf{v}$, $\mathbf{w} \in \mathrm{Bal}_n(R)$, then $\mathbf{v}' \circ \mathbf{w}'$ and $\mathbf{w}' \circ \mathbf{v}'$ both send $\mathbf{b}$ to $\mathbf{b} + \mathbf{v} + \mathbf{w}$. Since also $(\mathbf{v} + \mathbf{w})'(\mathbf{b}) = \mathbf{b} + \mathbf{v} + \mathbf{w}$, we have

$$\mathbf{v}' \circ \mathbf{w}' = (\mathbf{v} + \mathbf{w})' = \mathbf{w}' \circ \mathbf{v}'.$$

Thus the commutative law holds in $\mathrm{Mon}\langle \mathbf{v}' | \ \mathbf{v} \in \mathrm{Bal}_n(R)\rangle$, i.e., it is a *commutative monoid*. Since

$$\mathrm{Mon}(\mathcal{A}) = \mathrm{Mon}\langle f_{\mathbf{v}} | \ \mathbf{v} \in T\rangle \subseteq \mathrm{Mon}\langle \mathbf{v}' | \ \mathbf{v} \in \mathrm{Bal}_n(R)\rangle,$$

we conclude that $\mathrm{Mon}(\mathcal{A})$ is also a commutative monoid.

A further conclusion that may be drawn from the above equation is that the function determined by the assignment $\mathbf{v} \mapsto \mathbf{v}'$ is a *monoid homomorphism* from $\mathrm{Mon}\langle \mathbf{v} | \mathbf{v} \in T\rangle$ to $\mathrm{Mon}(\mathcal{A})$. Here $\mathrm{Mon}\langle \mathbf{v} | \mathbf{v} \in T\rangle$ denotes the submonoid of $\mathrm{Bal}_n(R)$ generated by all $\mathbf{v}$ in T. The term "homomorphism" in this context refers to the law $(\mathbf{v} + \mathbf{w})' = \mathbf{v}' \circ \mathbf{w}'$. Since $\mathbf{v}' = f_{\mathbf{v}}$ for $\mathbf{v} \in T$ and such functions $f_{\mathbf{v}}$ generate $\mathrm{Mon}(\mathcal{A})$, the function $\mathbf{v} \mapsto \mathbf{v}'$ is surjective. It is clearly injective, so it is bijective and hence is a *monoid isomorphism*. Summing up, we have a result which gives a simpler description of the monoid of an unbounded accounting system.

(6.2.2). *Let $\mathcal{A} = (A| \ T| \ \mathrm{Bal}_n(R))$ be an unbounded accounting system on n accounts over an ordered domain R. Then*

1. $\mathrm{Mon}(\mathcal{A})$ *is a commutative monoid;*

2. $\mathrm{Mon}(\mathcal{A}) \simeq \mathrm{Mon}\langle \mathbf{v} | \mathbf{v} \in T\rangle$ *and hence $\mathrm{Mon}(\mathcal{A})$ is isomorphic with a submonoid of $\mathrm{Bal}_n(R)$.*

Example (6.2.2).

A still less realistic type of accounting system is where there are restrictions on neither transactions nor balances, as in the accounting system

$$\mathcal{A} = (A| \ \mathrm{Bal}_n(R)| \ \mathrm{Bal}_n(R)).$$

In 4.2 this was called a free accounting system since it is devoid of restrictions, all transaction and balance vectors being allowable. In this case $T = \mathrm{Bal}_n(R)$ and 6.2.2 takes the following form.

Corollary. *If* $\mathcal{A} = (A|\ \mathrm{Bal}_n(R)|\ \mathrm{Bal}_n(R))$ *is a free accounting system, then* $\mathrm{Mon}(\mathcal{A}) \simeq \mathrm{Bal}_n(R)$.

Despite its impracticality, the free system on account set A has theoretical significance since it is the "largest" possible accounting system on A. It is clear from these examples that the presence of non-allowable balances complicates the structure of the monoid of an accounting system; for example, the monoid may be non-commutative.

Recall from 4.2 that a *feasible* transaction for an accounting system $\mathcal{A} = (A \mid T \mid B)$ is a transaction that is a composite of allowable transactions, i.e., a composite of transactions arising from the set T. In the case where $B = \mathrm{Bal}_n(R)$, i.e., the system is unbounded, the feasible transactions correspond exactly to the elements of the monoid $\mathrm{Mon}(\mathcal{A})$ by 6.2.1. This underlines the importance of the monoid of the system. We state this result next.

(6.2.3). *The monoid of an unbounded accounting system is the set of all feasible transactions for the system.*

Monoids provide a way of comparing different accounting systems on the same set of accounts. Intuitively one would want to compare two such systems by looking at the transactions that can be executed by each system. In 4.2 we defined two accounting systems with the same account set to be *equivalent* if every allowable transaction of one system is a feasible transaction of the other. Thus, if we disregard the possible occurrence of non-allowable balances, the effect of a sequence of allowable transactions in one system will be identical with that obtained from a similar sequence in the other system. On the basis of these remarks and 6.2.3, we can state:

(6.2.4). *Two unbounded accounting systems with the same account set are equivalent if and only if they have the same monoid.*

Groups and accounting systems

Consider an unbounded accounting system $\mathcal{A} = (A|\ T|\ \mathrm{Bal}_n(R))$. We have seen in 6.2.2 that $\mathrm{Mon}(\mathcal{A})$ is isomorphic with $\mathrm{Mon}\langle \mathbf{v}|\mathbf{v} \in T\rangle$ and that $\mathrm{Mon}(\mathcal{A})$ is a commutative monoid. Now suppose that the set of allowable transactions T has the property that $-\mathbf{v} \in T$ whenever $\mathbf{v} \in T$, i.e., T is closed with respect to forming negatives of its elements. Since $\mathbf{v}' \circ (-\mathbf{v}')$ and $(-\mathbf{v}') \circ \mathbf{v}'$ are both equal to the identity function,

$$(-\mathbf{v})' = (\mathbf{v}')^{-1},$$

which is equivalent to saying that *the inverse of an allowable transaction is allowable.*

What this means for the accounting system is that it is possible to reverse an allowable transaction, i.e., to correct a previously applied transaction. Thus the system has the capacity to correct errors. For this reason an accounting system $\mathcal{A} = (A|\,T|\,\mathrm{Bal}_n(R))$ is called *error correcting* if T is closed under forming negatives. We will discuss error correcting systems and related types of accounting systems in Chapter 7. The algebraic consequence of the error correcting property is that every element of $\mathrm{Mon}(\mathcal{A})$ has an inverse given by the formula

$$(\mathbf{v'}_1 \circ \mathbf{v'}_2 \circ \cdots \circ \mathbf{v'}_r)^{-1} = (-\mathbf{v}_1 - \mathbf{v}_2 - \cdots - \mathbf{v}_r)',$$

where $\mathbf{v}_i \in T$ and hence $-\mathbf{v}_i \in T$.

Thus we have a commutative monoid in which every element has an inverse, i.e., we have an abelian group. Hence we have:

(6.2.5). *If $\mathcal{A}$ is an unbounded, error correcting system, then* $\mathrm{Mon}(\mathcal{A})$ *is an abelian group.*

From the point of view of the algebraist, groups are preferable objects to work with since their structure is much better understood. It may be objected that in practice the inverse of an allowable transaction might not be allowable. For example, a transfer of funds from cash to an employee pension account would be routine, but a transfer in the other direction could be questionable. On the other hand, it is reasonable to suppose that an accounting system should have the ability to correct erroneous entries. Such errors could then be corrected by applying the inverse of an allowable transaction, which should therefore be allowable. Of course, a system with a built-in error correcting capability would need to be secure and come equipped with appropriate control mechanisms, features that will be considered in Chapter 9.

6.3. Accounting Systems as Automata II

In this section we introduce a second, more complex automaton to represent the operation of an accounting system. Like the previous automaton, this one keeps track of the transactions and balances of the system, but it does so in a different way, displaying the history of each account as part of a state of the system: in addition the output function is used to compute current balances and record the profit or loss of the system. In addition this automaton incorporates the *expiration times*, the earliest times that a transaction can be applied, and for this reason it will be called the *time enhanced* automaton of the system.

To help motivate the definition, let us consider the general situation of a company when an economic event affecting it occurs. This is illustrated in the schematic diagram below:

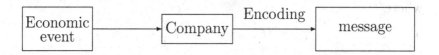

The economic event is assumed to affect the company through a *message*, which takes the form of a pair $(t, \mathbf{x})$ where t is the time of application of the transaction given by the balance vector $\mathbf{x}$ with entries $x_1, x_2, \ldots, x_n$ and n is the number of accounts.

Definition of the time enhanced automaton

To start things off, suppose that $\mathcal{A}$ is a free accounting system with n accounts

$$\mathcal{A} = (A \mid \mathrm{Bal}_n(R) \mid \mathrm{Bal}_n(R)),$$

where, as usual, R is an ordered domain. We will indicate later how freeness can be relaxed, but for the present we prefer to keep it as a simplifying assumption.

A *message of dimension n* is defined to be an ordered pair

$$(t, \mathbf{x}) \in \mathbb{Z} \times \mathrm{Bal}_n(R);$$

thus $t \in \mathbb{Z}, x_i \in R$ and of course $x_1 + x_2 + \cdots + x_n = 0$. The x_i are the entries of the balance vector which represents a transaction, while t is an integer specifying the earliest time of application of the transaction, which is called the *expiration* of the message.

Accounts and states of the automaton

The accounting system is assumed to have n accounts $a_1, a_2, \ldots,$ a_n, which are ordered in such a way that

$$a_{r+1}, a_{r+2}, \ldots, a_n$$

are the accounts that represent revenues and expenses.

A *state of the ith account* is defined to be an element of the set $\overline{\mathbb{Z} \times R}$, that is, a sequence of elements of $\mathbb{Z} \times R$

$$\overline{z}_i = z_{i1} z_{i2} \ldots z_{im(i)},$$

where

$$z_{ij} = (t_{ij}, x_{ij}), \ j = 1, 2, \ldots, m(i),$$

and $t_{ij} \in \mathbb{Z}, x_{ij} \in R$. In addition these entities are subject to the conditions

$$t_{i1} \leq t_{i2} \leq \cdots \leq t_{im(i)},$$

and

$$x_{ij} \neq 0 \ \text{ and } \ \sum_{i=1}^{n} \sum_{j=1}^{m(i)} x_{ij} = 0.$$

The significance of t_{ij} is that it is the expiration time for the jth transaction to affect the ith account. Notice that $m(i)$, the number of transactions affecting the ith account, depends on i. Intuitively we can think of $\overline{z}_i$ as recording the "history" of the ith account up to time $t_{im(i)}$.

The state of an account can be represented by an enhanced *T-diagram*, as illustrated in the figure below. The left hand column records the debits and the right column the credits; the difference from the T-diagrams defined in 3.1 is that in each case the expiration time is appended to the transaction amount.

Account a_i

$$
\begin{array}{c|c}
(t_{ij_1}, x_{ij_1}) & (t_{ij_{k(i)+1}}, x_{ij_{k(i)+1}}) \\
(t_{ij_2}, x_{ij_2}) & (t_{ij_{k(i)+2}}, x_{ij_{k(i)+2}}) \\
\vdots & \vdots \\
(t_{ij_{k(i)}}, x_{ij_{k(i)}}) & (t_{ij_{m(i)}}, x_{ij_{m(i)}})
\end{array}
$$

In the diagram it is assumed that $x_{ij} > 0$ and $j_1, j_2, \ldots, j_{m(i)}$ is a permutation of the integers $1, 2, \ldots, m(i)$ such that

$$
j_1 < j_2 < \cdots < j_{k(i)} \quad \text{and} \quad j_{k(i)+1} < j_{k(i)+2} < \cdots < j_{m(i)}.
$$

Since the left hand column records the debits on the account and the right hand column the credits, the value of the ith account at the latest time $t_{im(i)}$ will be increased by

$$
x_{ij_1} + x_{ij_2} + \cdots + x_{ij_{k(i)}} - x_{ij_{k(i)+1}} - x_{ij_{k(i)+2}} - \cdots - x_{ij_{m(i)}}
$$

if this is positive; should it be negative, there a corresponding decrease.

The operation of the time enhanced accounting system

We will now define the *time enhanced automaton*

$$
\mathcal{T}_A
$$

of the free accounting system $\mathcal{A}$ and explain how it functions. In the first place we have already defined a state of the ith account as an element of the set $\overline{\mathbb{Z} \times R}$ satisfying certain conditions. This allows us to define a *state of the automaton* $\mathcal{T}_A$ as a sequence of states of the n accounts, so that a typical state has the form

$$
\overline{z}_1 \overline{z}_2 \ldots \overline{z}_n,
$$

where

$$
\overline{z}_i = z_{i1} z_{i2} \ldots z_{im(i)} \quad \text{and} \quad z_{ij} = (t_{ij}, x_{ij}), \quad t_{ij} \in \mathbb{Z}, \ x_{ij} \in R.
$$

Here it is understood that the previously stated conditions must be satisfied, i.e.,

$$
t_{i1} \leq t_{i2} \leq \cdots \leq t_{im(i)},
$$

and

$$x_{ij} \neq 0 \text{ and } \sum_{i=1}^{n} \sum_{j=1}^{m(i)} x_{ij} = 0.$$

The set of states of the automaton $\mathcal{T}_A$ is to be a set Z where

$$Z \subseteq (\overline{\mathbb{Z} \times R})^n.$$

Next the inputs are messages, i.e., elements of the set $\mathbb{Z} \times \text{Bal}_n(R)$, a typical one being of the form $(t, \mathbf{x})$. Thus the input set for $\mathcal{T}_A$ is

$$X = \mathbb{Z} \times \text{Bal}_n(R).$$

The next state function

$$\delta : Z \times X \longrightarrow Z,$$

is defined by the rules that follow:

$$\delta(\overline{z}_1 \overline{z}_2 \ldots \overline{z}_n, (t, \mathbf{x})) = \overline{z}_1 \overline{z}_2 \ldots \overline{z}_n$$

if $t < t_{ij}$ for some $j = 1, 2, \ldots, m(i)$, $i = 1, 2, \ldots, n$; on the other hand,

$$\delta(\overline{z}_1 \overline{z}_2 \ldots \overline{z}_n, (t, \mathbf{x})) = \overline{z}_1(t, x_1) \overline{z}_2(t, x_2) \ldots \overline{z}_n(t, x_n),$$

if $t \geq t_{ij}$ for all $j = 1, 2, \ldots, m(i)$, $i = 1, 2, \ldots, n$: observe here that $\overline{z}_i(t, x_i)$ is the concatenation of $\overline{z}_i$ and (t, x_i), except that if $x_i = 0$, the pair (t, x_i) is to be omitted from the sequence.

What is happening here is that the state of the automaton will not change unless $t \geq t_{im(i)}$ for $i = 1, 2, \ldots, n$, i.e., t does not precede any expiration time, in which event the new state is obtained by adjoining the pair (t, x_i) to $\overline{z}_i$ unless $x_i = 0$.

Example (6.3.1).

Let us see how the function δ acts in a particular case. Suppose that an accounting system has three accounts a_1, a_2, a_3, the third account being an income or expense account. Assume that at some instant the states, i.e., histories, $\overline{z}_1, \overline{z}_2, \overline{z}_3$ of the three accounts are given by the T-diagrams below.

Account a_1		Account a_2	Account a_3	
$(t_1, 400)$	$(t_3, 100)$	$(t_1, 100)$	$(t_3, 100)$	$(t_1, 300)$
$(t_2, 600)$		$(t_2, 600)$		$(t_4, 200)$
$(t_4, 900)$		$(t_4, 700)$		

The understanding here is that $t_1 < t_2 < t_3 < t_4$. These diagrams describe the state of the automaton at time t_4 as $\overline{z}_1 \overline{z}_2 \overline{z}_3$ where, for example, $\overline{z}_1 = (t_1, 400)(t_2, 600)(t_3, -100)(t_4, 900)$.

Now suppose that the input $(t_5, (0, -500, 500))$ is read by the automaton where $t_5 \geq t_4$. Then it follows from the definition of the function δ that the next state of the automaton is given by the T-diagrams

Account a_1		Account a_2	Account a_3	
$(t_1, 400)$	$(t_3, 100)$	$(t_1, 100)$	$(t_3, 100)$	$(t_1, 300)$
$(t_2, 600)$		$(t_2, 600)$	$(t_5, 500)$	$(t_4, 200)$
$(t_4, 900)$		$(t_4, 700)$		
		$(t_5, 500)$		

Keep in mind that the left hand column is for debits and the right hand column for credits. Thus $(t_5, 500)$ has been adjoined to the credit column of the second account and to the debit column of the third account. On the other hand, since the input has a 0 entry for the first account, no pairs are added to its T-diagram.

To complete the description of the automaton $\mathcal{T}_A$, the output set and output function must be assigned. We take as the output set

$$Y = (\mathbb{Z} \times \text{Bal}_{r+1}(R)) \cup \{e\}$$

where e is an error message: recall that r is the number of accounts which are not revenue or expense accounts. Then the output function $\lambda : Z \times X \longrightarrow Y$ is defined by the following rule: λ sends $(\overline{z}_1 \overline{z}_2 \ldots \overline{z}_n, (t, \mathbf{x}))$ to $(t, \overline{\mathbf{x}})$ where $\overline{\mathbf{x}}$ has components

$$x_1 + \sum_{j=1}^{m(1)} x_{1j}, \ x_2 + \sum_{j=1}^{m(2)} x_{2j}, \ \ldots, \ x_r + \sum_{j=1}^{m(r)} x_{rj}, \ \sum_{i=r+1}^{n} \left(x_i + \sum_{j=1}^{m(i)} x_{ij} \right),$$

except that if $t < t_{ij}$ for some i, j, then the value of the output function is the error message e.

What the output function does here, assuming that no error message is generated, is first to record the time of application of the transaction; then for each of the accounts $a_1, a_2, \ldots, a_r$ it sums the values of all the transactions, thereby giving the net change in value of these accounts. It also totals the transaction values for all the revenue and expense accounts $a_{r+1}, \ldots, a_n$ and places this sum as the $(r + 1)$th entry of the output; clearly this represents the net income of the system after the transaction has been processed.

Notice that the partition

$$A = \{a_1\} \cup \{a_2\} \cup \cdots \cup \{a_r\} \cup \{a_{r+1}, \ldots, a_n\}$$

determines an equivalence relation E on A, and hence the quotient system $\overline{A} = A/E$. There is a natural epimorphism from A to $\overline{A}$ which was described in 5.2. The $(r + 1)$th entry of

$$\lambda((\overline{z}_1\overline{z}_2 \ldots \overline{z}_n, (t, \mathbf{x})))$$

represents the value of the $(r + 1)$th account of the quotient system $\overline{A}$, an account that could be called the profit or loss account for A: however, strictly speaking it is an account of $\overline{A}$, not A.

The automaton just constructed,

$$T_A = (Z, X, Y, \delta, \lambda),$$

is the time enhanced automaton of the accounting system A. Next we give an example to illustrate the computation of the output function of the automaton T_A.

Example (6.3.2).
Consider again the accounting system in Example 6.3.1 as specified by the T-diagrams

Account 1		Account 2	Account 3	
$(t_1, 400)$	$(t_3, 100)$	$(t_1, 100)$	$(t_3, 100)$	$(t_1, 300)$
$(t_2, 600)$		$(t_2, 600)$		$(t_4, 200)$
$(t_4, 900)$		$(t_4, 700)$		

As before assume that the message $(t_5, (0, -500, 500))$ is received, where $t_4 < t_5$. Now we compute the output as specified in the definition of the output function, keeping in mind that a_3 is a revenue or expense account. The output is $(t_5, (1800, -1900, 100))$. Thus over the entire history of the system account a_1 has final value 1800 and account a_2 has value -1900, while the third entry tells us there has been a net loss to the system of 100.

Additional remarks on the time enhanced automaton

We maintain the notation already established for the time enhanced automaton $\mathcal{T}_A$ of an unbounded accounting system $\mathcal{A}$.

1. The output $\lambda(\bar{z}_1\bar{z}_2\ldots\bar{z}_n, (t, \mathbf{x}))$ is the *balance sheet at time t*. Its components give the balances of non-revenue or expense accounts, and the net profit or loss of the system. Specifically, the ith non-time component of $\lambda(\bar{z}_1\bar{z}_2\ldots\bar{z}_n, (t, (\mathbf{x})))$, $i = 1, 2, \ldots, r$, is the balance of the ith account at time t, while the $(r+1)$th component is the net profit or loss of the system.

2. If a name is assigned to each component of the set of states Z, the set of all these denominations is called an *account plan*: its components represent the history of each account.

3. The automaton $\mathcal{T}_A$ can also compute the final values of the accounts. The input set $X = \mathbb{Z} \times \text{Bal}_n(R)$ can be extended to the free monoid $\overline{X}$ on X, as described in 6.1. Thus the input is a finite sequence of messages applied in succession. Suppose that the initial state of $\mathcal{T}_A$ was

$$z_1^0 z_2^0 \ldots z_n^0,$$

where

$$z_i^0 = (t_0, x_i^0), \quad i = 1, 2, \ldots, r$$

and

$$z_i^0 = (t_0, 0), \quad i = r+1, r+2, \ldots, n.$$

Thus it is assumed that the revenue and expense accounts had zero balances initially, as would normally be the case. Suppose that a sequence of m inputs, i.e., messages, is applied,

$$(t_i, \mathbf{x}_i), \quad i = 1, 2, \ldots, m,$$

where $t_0 < t_1 < t_2 < \cdots < t_m$. Then the value of the output function after the sequence of inputs has been processed, i.e.,

$$\overline{\lambda}(z_1^0 z_2^0 \ldots z_n^0, (t_1, \mathbf{x}_1) \ldots (t_m, \mathbf{x}_m))$$

is the *final balance sheet of the accounting system*; of course the value of the output function is computed recursively.

4. Usually the output function λ is not applied each time a message is received, but only at certain moments in the life of the company, for example, at the end of the year or the quarter. The process of calculating the value of λ is called the *regularization*.

5. The definition of the time enhanced automaton has been formulated for free accounting systems. However, it is not difficult to see how to modify the definition for a general accounting system. It is necessary to screen out non-allowable transactions or balances by adjusting appropriately the definitions of the next state and output functions, as was done for the automaton of 6.2. This will require that additional error messages be adjoined to the output set Y.

6. Just as for any automaton, we can define various equivalence relations for the time enhanced automaton. A relation is introduced on the set of states by defining two states to be equivalent if, on receiving the same input, they produce the same output. Thus two states are equivalent in this sense if they always lead to the same balance sheet. Dually we can define equivalence of two messages, i.e., inputs, if, starting from the same state, they always lead to the same output, i.e., the same balance sheet.

Chapter Seven

Accounting Systems with Restricted Transactions

7.1. An Overview of Special Systems

The previous chapters have seen the construction of our basic algebraic model, which is able to simulate many of the operations of the double-entry accounting system. However, it is clear that this model is far from being a realistic system. In the first place there are still facets of accounting not represented in the model, control of transactions being a prominent example. But it is also evident that the model is too general and includes systems that could never be implemented by any computer system. With this in mind, we shall consider in this chapter what restrictions can reasonably be imposed on the allowable transactions of an accounting system. The object is, of course, to arrive at a system which is closer to reality and stands a chance of machine implementation.

One natural restriction is to require that there be only finitely many specific allowable transactions. In practice these might represent fixed regular payments or receipts and so they would be finite in number. Thus we are led to a class of accounting systems called *finitely specified systems*, in which an allowable transaction is either one of an approved type or else one of a finite number of specific transactions. More generally, we might accept accounting systems that are equivalent to finitely specified systems, meaning that they have the same sets of feasible transactions. Such systems will be called *finitely specifiable*. These may be thought of as the widest class of unbounded systems that could conceivably be machine implemented. Incidentally the algorithms described in Chapter 8 make a convincing case for this statement.

From the algebraic point of view there are several subclasses of finitely specifiable systems that stand out. A full discussion of these types follows in 7.2, together with an assessment of their interrelationships and their significance for accounting. Prominent among the finitely specifiable systems are systems that are equivalent to systems specified by simple transactions only; these are termed *simple systems* and they are much easier to analyze. They are treated in some detail in 7.3. It seems a reasonable assessment that many real life systems will be simple or at least nearly so.

Another significant class of finitely specifiable systems is the class of systems in which the inverse of every feasible transaction is feasible. These are called *inverse systems* and they are distinguished by the property that their monoids are groups. A closely related class of systems of particular significance for accounting consists of systems which are error correcting. In these the inverse of an allowable transaction is allowable, which makes it possible to correct erroneously entered transactions. This is a feature one would expect to find in any practical system, subject of course to appropriate control mechanisms.

Throughout this chapter we assume that all accounting systems are over the ring of integers $\mathbb{Z}$ (although one could work to some extent with systems over finitely generated ordered domains). In addition we will generally assume that all accounting systems are unbounded, i.e., every balance vector is allowable in the system. The main reason for this is that the monoid of such a system is much easier to understand, being isomorphic with a submonoid of the monoid $\mathrm{Bal}_n(\mathbb{Z})$ where n is the number of accounts – see 6.2.2. Also the relation between a system and its feasible digraph is much closer for unbounded systems. Having said this, we acknowledge that many of the concepts introduced still make sense for general accounting systems.

7.2. Finitely Specifiable Accounting Systems

Consider an unbounded accounting system on n accounts

$$\mathcal{A} = (A \mid T_0, T_1 \mid \mathrm{Bal}_n(\mathbb{Z})),$$

so that all balance vectors are allowable: for brevity we will write

$$\mathcal{A} = (A \mid T_0, T_1).$$

Here T_0 is the set of transaction types that are allowable and T_1 is the set of specific allowable transactions. Now T_0 is certainly finite: indeed $|T_0| \le 3^n - 2^{n+1} + 2$ by 3.2.1. On the other hand, the set T_1 could be infinite: at least the definition in Chapter 4 leaves this possibility open. In practice, T_1 is likely to be a fairly small set, consisting of regular operations such as repayments on a mortgage, interest on a bank loan or insurance premiums. It therefore seems a reasonable assumption that T_1 is finite.

With these remarks in mind, we define an unbounded accounting system $\mathcal{A}$ to be *finitely specified* if $\mathcal{A} = (A \mid T_0, T_1)$ where the set T_1 is finite. For example, we could form a finitely specified system by taking as allowable transaction vectors all the elementary vectors $\mathbf{e}(i, j)$. A useful feature of a finitely specified accounting system is that it can be completely described by a partitioned matrix. In this matrix the first set of columns are the allowable types of transaction vector in T_0, which therefore have entries 0, $+$ or $-$, while the remaining columns list the specific allowable transaction vectors in T_1. Thus the matrix has size $n \times (|T_0| + |T_1|)$. The representation of a finitely specified accounting system is therefore accomplished by a single matrix.

More generally, an accounting system $\mathcal{A} = (A \mid T_0, T_1)$ is said to be *finitely specifiable* if it is equivalent to a finitely specified system $\overline{\mathcal{A}} = (A \mid \overline{T}_0, \overline{T}_1)$: here of course $\overline{T}_1$ is finite, but T_1 might be infinite. Finitely specifiable systems can be characterized in terms of their monoids.

(7.2.1). *An unbounded accounting system $\mathcal{A}$ is finitely specifiable if and only if $\mathrm{Mon}(\mathcal{A})$ is generated by finitely many elements together with all elements of certain types.*

Proof
Suppose first that $\mathcal{A}$ is finitely specifiable. Then it is equivalent to a finitely specified system $\overline{\mathcal{A}} = (A \mid \overline{T}_0, \overline{T}_1)$ and $\mathrm{Mon}(\mathcal{A}) = \mathrm{Mon}(\overline{\mathcal{A}})$; thus $\mathrm{Mon}(\mathcal{A})$ is generated by the finite set $\overline{T}_1$, together with all vectors of types in $\overline{T}_0$.

Conversely, assume that $\mathrm{Mon}(\mathcal{A})$ is generated by a finite set T_1 and all vectors of some set of types T_0. Define $\overline{\mathcal{A}} = (A \mid T_0, T_1)$, which is a finitely specified system. Since $\mathrm{Mon}(\mathcal{A}) = \mathrm{Mon}(\overline{\mathcal{A}})$, we see that $\mathcal{A}$ is equivalent to $\overline{\mathcal{A}}$, so that $\mathcal{A}$ is finitely specifiable. $\square$

While it is obvious that every finitely specified system is finitely specifiable, *the converse is false*: for example, the free system $\mathcal{A} = (A|\,\mathrm{Bal}_n(\mathbb{Z}))$ is finitely specifiable since $\mathrm{Mon}(\mathcal{A}) = \mathrm{Bal}_n(\mathbb{Z})$ is generated by the $n(n-1)$ elementary vectors $\mathbf{e}(i,j)$, but it is not finitely specified.

Example (7.2.1).

There are accounting systems that are not finitely specifiable.

Consider the unbounded system $\mathcal{A}$ with three accounts and allowable transactions

$$\begin{bmatrix} 1 \\ 1+a \\ -2-a \end{bmatrix}, \ a > 0.$$

Each non-zero feasible transaction, i.e., element of $\mathrm{Mon}(\mathcal{A})$, must clearly be of type

$$\begin{bmatrix} + \\ + \\ - \end{bmatrix},$$

yet not every vector of this type is in $\mathrm{Mon}(\mathcal{A})$. Indeed, if $\begin{bmatrix} 1 \\ 1 \\ -2 \end{bmatrix}$ were feasible, there would be an expression

$$\begin{bmatrix} 1 \\ 1 \\ -2 \end{bmatrix} = \ell_1 \begin{bmatrix} 1 \\ 1+a_1 \\ -2-a_1 \end{bmatrix} + \cdots + \ell_k \begin{bmatrix} 1 \\ 1+a_k \\ -2-a_k \end{bmatrix}$$

with $\ell_i, a_i > 0$. On inspecting the top rows of the column vectors, we see that $1 = \ell_1 + \cdots + \ell_k$, which implies that $k = 1 = \ell_1$; but then a glance at the second rows reveals the contradiction $1 = 1 + a_1$.

Now suppose that $\mathcal{A}$ is finitely specifiable. Then, since in $\mathcal{A}$ not all vectors of any one type are feasible, $\mathrm{Mon}(\mathcal{A})$ must be generated by finitely many allowable vectors, say by $\begin{bmatrix} 1 \\ 1+a_i \\ -2-a_i \end{bmatrix}$, $i = 1, 2, \ldots, m$. However, if we choose a to be larger than all of $a_1, \ldots, a_m$, it is easy to see that $\begin{bmatrix} 1 \\ 1+a \\ -2-a \end{bmatrix}$ cannot be written in the form

$$\sum_{i=1}^{m} \ell_i \begin{bmatrix} 1 \\ 1 + a_i \\ -2 - a_i \end{bmatrix}$$

where $\ell_i \geq 0$. Consequently the system $\mathcal{A}$ cannot be finitely specifiable.

Next we introduce some special types of finitely specifiable accounting systems by further restricting their monoids.

Finitely generated accounting systems

Following a well-known algebraic path, let us call an (unbounded) accounting system

$$\mathcal{A} = (A \mid T)$$

finitely generated if $M = \text{Mon}(\mathcal{A})$ can be generated as a monoid by some finite subset $\{\mathbf{v}(1), \ldots, \mathbf{v}(m)\}$. This means that every element of M, i.e., every feasible transaction vector, is expressible in the form

$$\ell_1 \mathbf{v}(1) + \cdots + \ell_m \mathbf{v}(m)$$

with ℓ_i a non-negative integer. Clearly $\mathcal{A}$ is finitely specifiable since it is equivalent to the system whose allowable transactions are $\mathbf{v}(1), \ldots, \mathbf{v}(m)$. Notice that $\mathbf{v}(1), \ldots, \mathbf{v}(m)$ are feasible in $\mathcal{A}$, but are not necessarily allowable. However, each $\mathbf{v}(i)$ is expressible in terms of finitely many elements of T. Thus, while T itself might be infinite, it has a finite subset T_f such that each $\mathbf{v}(i)$ is a linear combination of vectors in T_f with non-negative integral coefficients. Thus $M = \text{Mon}(\mathcal{A}) = \text{Mon}\langle T_f \rangle$. Now let $\mathcal{A}_f$ be the finitely generated system $(A \mid T_f)$; then $\mathcal{A}$ and $\mathcal{A}_f$ have the same monoid M, so they are equivalent. This conclusion is stated as:

(7.2.2). *Let $\mathcal{A} = (A \mid T)$ be a finitely generated unbounded accounting system. Then there is a finite subset T_f of T such that $\mathcal{A}$ is equivalent to the finitely specified system $\mathcal{A}_f = (A \mid T_f)$. Thus every finitely generated system is finitely specifiable.*

Simple accounting systems

We recall from Chapter 2 that a balance vector of the form $a\mathbf{e}(i,j)$, $(a > 0)$, is said to be *simple*. Here $\mathbf{e}(i,j)$ is the elementary transaction vector with ith entry $+1$ and jth entry -1. An accounting system $\mathcal{A}$ will be called *simple* if its monoid can be generated by simple transaction vectors. Thus each allowable transaction can be expressed as a sum of feasible simple transactions. In practice many allowable transactions in an accounting system will be simple, which suggests that real-life systems may often be simple. However, the definition of a simple system requires only that the monoid be generated by simple transaction vectors; these need not be allowable, although they must of course be feasible.

A special type of simple system occurs when the feasible monoid is generated by elementary transaction vectors: such a system is called *elementary*. For an accounting system with n accounts, the total number of elementary transactions is $n(n-1)$; thus the monoid can be generated by $n(n-1)$ or fewer elements. Hence every *elementary system is finitely generated*. In fact this conclusion can be strengthened.

(7.2.3). *Every simple unbounded accounting system is finitely generated.*

The proof of this result rests on a property of the submonoids of the monoid $\mathbb{N}$ of natural numbers, namely every submonoid is finitely generated. We will give a proof of this result, but first an auxiliary lemma is necessary.

(7.2.4). *Let $a_1, a_2, \ldots, a_k$ be positive integers which are relatively prime. Then there exists $x_0 \in \mathbb{N}$ such that every integer $x \geq x_0$ belongs to $\mathrm{Mon}\langle a_1, a_2, \ldots, a_k \rangle$: furthermore x_0 can be computed from the a_i.*

Proof
Write $S = \mathrm{Mon}\langle a_1, \ldots, a_k \rangle$. Since $a_1, \ldots, a_k$ are relatively prime, there are integers $r_1, \ldots, r_k$ such that

$$1 = r_1 a_1 + \cdots + r_k a_k,$$

where we can suppose that $r_i < 0$ for $i = 1, 2, \ldots, \ell < k$ and $r_i \geq 0$ for $i = \ell+1, \ell+2, \ldots, k$. Put $s_i = -(a_k - 1)r_i$ for $i = 1, \ldots, \ell$, noting that $s_i \geq 0$. Hence $x_0 = s_1 a_1 + \cdots + s_\ell a_\ell \in S$. For $0 \leq j \leq a_k - 1$ we have

$$x_0 + j = s_1 a_1 + \cdots + s_\ell a_\ell + j r_1 a_1 + \cdots + j r_k a_k,$$

which equals

$$(s_1 + j r_1) a_1 + \cdots + (s_\ell + j r_\ell) a_\ell + j r_{\ell+1} a_{\ell+1} + \cdots + j r_k a_k.$$

If $1 \leq i \leq \ell$ and $0 \leq j \leq a_k - 1$, then

$$s_i + j r_i = -(a_k - 1) r_i + j r_i = -r_i(a_k - 1 - j) \geq 0$$

since $r_i < 0$. Hence $x_0 + j \in S$ for $j = 0, 1, \ldots, a_k - 1$.

Now suppose that there exists an $x > x_0$ in $\mathbb{N}\backslash S$; then we can assume that x has been chosen minimal with these properties. If $x - a_k \geq x_0$, then $x - a_k \in S$ by minimality of x, so that $x = (x - a_k) + a_k \in S$. By this contradiction $x - a_k < x_0$ and thus $x_0 < x \leq x_0 + (a_k - 1)$. But then $x \in S$ since x is of the form $x_0 + j$ for some j satisfying $0 < j \leq a_k - 1$, a contradiction which shows that all integers $x \geq x_0$ belong to S. $\qquad \square$

The crucial result needed to establish 7.2.3 can now be proved.

(7.2.5). *If S is a submonoid of $\mathbb{N}$, then it can be finitely generated as a monoid.*

Proof
Assume that S is not finitely generated, which implies that $S \neq 0$. Let a_1 be any non-zero element of S. We show how to construct a sequence of elements $a_1, a_2 \ldots$ of S such that, if d_i is the greatest common divisor of $a_1, a_2 \ldots, a_i$, then $d_i > d_{i+1}$. Assume that the sequence has been constructed as far as a_k. Since $\frac{a_1}{d_k}, \frac{a_2}{d_k}, \ldots, \frac{a_k}{d_k}$ are relatively prime, 7.2.4 shows that

$$\mathrm{Mon}\langle d_k \rangle \backslash \mathrm{Mon}\langle a_1, a_2, \ldots, a_k \rangle$$

is finite. Suppose that $S \subseteq \mathrm{Mon}\langle d_k \rangle$; then $S \backslash \mathrm{Mon}\langle a_1, a_2, \ldots, a_k \rangle$ is finite, which leads to the contradiction that S is finitely generated. Hence there exists an $a_{k+1} \in S$ which is not divisible by d_k. Then, writing d_{k+1} for the greatest common divisor of $a_1, a_2, \ldots, a_{k+1}$, we have $d_k > d_{k+1}$, so the construction has been effected. Since the d_i are positive integers, there must exist a k for which $d_k = 1$. Then

$S \setminus \text{Mon}\langle a_1, a_2, \ldots, a_k \rangle$ is finite by 7.2.4. However, this implies that S is finitely generated, which is a contradiction. $\square$

Proof of 7.2.3

Assume that $\mathcal{A}$ is a simple unbounded system with monoid M. Thus M is generated by certain simple transactions. For each $i \neq j$ let M_{ij} be the set of integers $a \geq 0$ such that $a\mathbf{e}(i,j) \in M$. If $a, b \in M_{ij}$, then $a + b \in M_{ij}$, which shows that M_{ij} is a submonoid of $\mathbb{N}$ (possibly zero). We now apply 7.2.5 to show that M_{ij} can be generated by finitely many of its elements, say $a_{ij}^{(k)}$, $k = 1, 2, \ldots, \ell(i,j)$. Then the elements $a_{ij}^{(k)}\mathbf{e}(i,j)$, $k = 1, 2, \ldots, \ell(i,j)$, $1 \leq i, j \leq n$, generate the monoid M and the number of these elements is finite, so M is finitely generated. $\square$

Corollary. *Every unbounded accounting system with two accounts is finitely generated.*

Proof

An accounting system with two accounts is simple, so the result follows at once from 7.2.3. $\square$

On the other hand, there are accounting systems with three accounts which are not finitely generated – see Example 7.2.3(ii) below.

Hereditary accounting systems

In Chapter 3 a partial ordering of n-transaction types was introduced which classifies transactions according to their complexity. In this ordering, if $\mathbf{t}$ and $\mathbf{s}$ are two type vectors of equal size, so that each of their entries is 0, $+$ or $-$, then $\mathbf{t} \leq \mathbf{s}$ means that for each i either $t_i = 0$ or $t_i = s_i$. It seems *a priori* to be a plausible hypothesis that if a transaction vector $\mathbf{v}$ is feasible and $\mathbf{u}$ is a transaction of the same or an earlier type in the ordering, i.e., $\text{type}(\mathbf{u}) \leq \text{type}(\mathbf{v})$, then $\mathbf{u}$ should also be feasible.

With this in mind, let us call an unbounded system $\mathcal{A}$ *hereditary* if, whenever $\mathbf{v}$ is feasible in $\mathcal{A}$ and $\text{type}(\mathbf{u}) \leq \text{type}(\mathbf{v})$, then $\mathbf{u}$ is also feasible in $\mathcal{A}$. It will emerge from the analysis in 7.3 that hereditary systems are very special and are in fact completely determined by their feasible digraphs.

Example (7.2.2).

Suppose that the vector

$$\mathbf{v} = \begin{bmatrix} 400 \\ -100 \\ -200 \\ -100 \end{bmatrix}$$

is feasible in a hereditary system $\mathcal{A}$ with four accounts. Then all vectors with type preceding

$$\text{type}(\mathbf{v}) = \begin{bmatrix} + \\ - \\ - \\ - \end{bmatrix}$$

are feasible. In particular $\mathbf{e}(1,2)$, $\mathbf{e}(1,3)$ and $\mathbf{e}(1,4)$ are feasible. Now observe that $\mathbf{v}$ can be expressed in terms of these elementary feasible vectors, indeed

$$\mathbf{v} = 100\mathbf{e}(1,2) + 200\mathbf{e}(1,3) + 100\mathbf{e}(1,4).$$

Hence $\mathbf{e}(1,2), \mathbf{e}(1,3), \mathbf{e}(1,4)$ generate the monoid of $\mathcal{A}$, which is therefore an elementary system. In fact this statement is true in any hereditary system.

(7.2.6). *Every hereditary, unbounded accounting system $\mathcal{A}$ is elementary.*

Proof
Let $\mathbf{v}$ be a non-zero feasible transaction vector for $\mathcal{A}$. It is necessary to show that $\mathbf{v}$ is a sum of elementary feasible vectors of $\mathcal{A}$. The proof is by induction on the number $m \geq 2$ of non-zero entries of $\mathbf{v}$.

Since $\mathbf{v} \neq \mathbf{0}$, it must have a positive entry v_r and a negative entry v_s. A new transaction vector $\mathbf{u}$ is defined by the rule

$$\mathbf{u} = \begin{cases} \mathbf{v} + v_s\,\mathbf{e}(r,s) & \text{if } v_r + v_s > 0, \\ \mathbf{v} - v_r\,\mathbf{e}(r,s) & \text{if } v_r + v_s \leq 0. \end{cases}$$

Now observe that $\text{type}(\mathbf{u}) \leq \text{type}(\mathbf{v})$, so that $\mathbf{u}$ is feasible by the hereditary property, and also that the number of non-zero entries of $\mathbf{u}$ is at most $m - 1$. Therefore $\mathbf{u}$ is a sum of elementary feasible vectors by the induction hypothesis. Next $\text{type}(\mathbf{e}(r,s)) \leq \text{type}(\mathbf{v})$, so that $\mathbf{e}(r,s)$ is feasible in $\mathcal{A}$. Finally $\mathbf{v} = \mathbf{u} + (-v_s)\mathbf{e}(r,s)$ or $\mathbf{u} + v_r\,\mathbf{e}(r,s)$ and $v_r > 0,\ -v_s > 0$; hence $\mathbf{v}$ is a sum of elementary feasible vectors, as claimed.   □

Type-complete accounting systems

Another special type of unbounded system occurs when, given that a transaction vector is feasible, it follows that all vectors of the same type are feasible. Such a system will be called *type-complete*. Notice that hereditary systems are type-complete, but the latter is evidently a weaker property.

A still weaker property than type-completeness is purity. Here an unbounded system is said to be *pure* if its monoid is generated by all the transaction vectors of certain given types. Thus a pure system is equivalent to a finitely specified system in which there are no specific allowable transactions, only allowable types: therefore a pure system is finitely specifiable. Notice that elementary accounting systems are pure since their monoids are generated by certain elementary vectors $\mathbf{e}(i,j)$ and hence contain all simple vectors of the same type.

It is difficult to assess how often a practical accounting system would be pure or type-complete. These properties are mentioned here primarily because they appear natural from the algebraic standpoint.

Inverse accounting systems

An accounting system $\mathcal{A}$ is called an *inverse system* if the inverse of a feasible transaction, or equivalently the negative of the corresponding transaction vector, is always feasible. This is equivalent to saying that $\text{Mon}(\mathcal{A})$ is a subgroup of the abelian group $\text{Bal}_n(\mathbb{Z})$. In an inverse system the negative of an allowable transaction might not be allowable, but it is certainly feasible. Recall from 6.2 that an (unbounded) accounting system is called *error correcting* if the negative of every allowable transaction vector is allowable. Every error correcting system $\mathcal{A}$ is an inverse system. To see this, let $\mathbf{v}$ be a feasible transaction vector of $\mathcal{A}$, so that there is an expression $\mathbf{v} = \mathbf{v}_1 + \cdots + \mathbf{v}_k$, where the $\mathbf{v}_i$ are allowable. Then $-\mathbf{v}_i$ is allowable since $\mathcal{A}$ is error correcting. Hence $-\mathbf{v} = (-\mathbf{v}_1) + \cdots + (-\mathbf{v}_k)$ is feasible and thus $\mathcal{A}$ is an inverse system.

Conversely, if $\mathcal{A} = (A \mid T)$ is an inverse system, we can modify the system by adjoining the negatives of all the allowable transactions of

$\mathcal{A}$, thereby obtaining an error correcting system $\overline{\mathcal{A}} = (A \mid T \cup (-T))$ where $-T$ denotes the set $\{-\mathbf{v} \mid \mathbf{v} \in T\}$. Clearly $\mathcal{A}$ is equivalent to $\overline{\mathcal{A}}$.

Summing up the discussion, we have:

(7.2.7). *Let $\mathcal{A}$ be an unbounded accounting system. Then the following hold.*

1. *If $\mathcal{A}$ is error correcting, then it is an inverse system.*

2. *If $\mathcal{A}$ is an inverse system, then it is equivalent to an error correcting system.*

The question arises as to how inverse systems are related to the other types of system described in this section.

(7.2.8). *If $\mathcal{A}$ is an inverse, unbounded accounting system on n accounts, then the feasible monoid of $\mathcal{A}$ can be generated by at most $2(n-1)$ elements. Thus $\mathcal{A}$ is a finitely generated system.*

Proof

Let M denote the monoid of $\mathcal{A}$, so that M is a subgroup of $\mathrm{Bal}_n(\mathbb{Z})$. Now $\mathrm{Bal}_n(\mathbb{Z})$ is a free abelian group of rank $n-1$ by 2.3.1. It is a well-known fact about free abelian groups that M can be generated *as a group* by at most $n-1$ elements – see [7] or [8]. The group generators and their negatives will then generate M *as a monoid*. Therefore M can be generated as a monoid by at most $2(n-1)$ of its elements. $\qquad\qquad\square$

There is good reason to believe that a real life accounting system will have the ability to correct errors. Inevitably in day-to-day operations errors will arise through the entering of erroneous data or system malfunctions. If an erroneous transaction has been applied to the accounting system, it can be corrected by applying the inverse transaction, i.e., the negative of the transaction vector. (Notice that even if balance restrictions are present, the correcting transactions will restore the original balance vector). Of course in an error-correcting system there is increased risk of misuse, but this can be reduced by introducing a control mechanism, as described in Chapter 9.

Diagram of classes of accounting systems

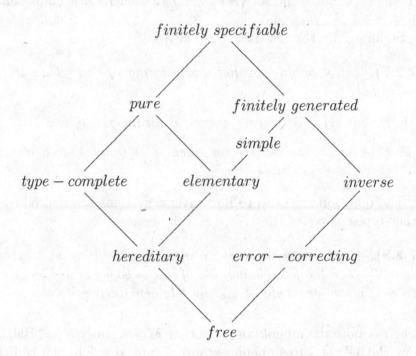

Since many classes of unbounded accounting systems have been introduced in this section, it is useful to display these classes by means of an inclusion diagram as above. Here the larger classes of systems appear higher up in the diagram, while inclusions between classes are indicated by sequences of upward directed lines.

Next we will show that there are no inclusions between the ten classes of accounting systems discussed above other than those displayed in the diagram. This is demonstrated by a series of simple examples.

Example (7.2.3).

(i) *Error-correcting does not imply either pure or simple.*

This is shown by the 3-account system with allowable transaction vectors

$$\begin{bmatrix} 1 \\ -2 \\ 1 \end{bmatrix} \quad \text{and} \quad \begin{bmatrix} -1 \\ 2 \\ -1 \end{bmatrix}.$$

Here transaction vectors of only two types are feasible, but not all vectors of these types are feasible, so the system is not pure. Also it is not simple since no simple transactions are feasible.

(ii) *Type-complete does not imply finitely generated.*

Consider the 3-account system $\mathcal{A}$ whose allowable transaction vectors are all those of type $\begin{bmatrix} + \\ + \\ - \end{bmatrix}$. Then $\mathrm{Mon}(\mathcal{A})$ is not finitely generated. Indeed, suppose it is generated by

$$\begin{bmatrix} a_i \\ b_i \\ -a_i - b_i \end{bmatrix}, \ i = 1, \ldots, k,$$

where $a_i, b_i > 0$. Let $x, y > 0$; then the vector which has entries x, y, $-x - y$ belongs to $\mathrm{Mon}(\mathcal{A})$, so there exist $\ell_i \geq 0$ such that

$$\begin{cases} x = \ell_1 a_1 + \cdots + \ell_k a_k \\ y = \ell_1 b_1 + \cdots + \ell_k b_k \end{cases}$$

Choose x to be the smallest of $a_1, \ldots, a_k$, say a_1, and $y = 1 + b_1$; then $k = 1$ and $\ell_1 = 1$. But then a contradiction ensues since $y \neq b_1$.

(iii) *Hereditary does not imply inverse.*

This is shown by the 2-account system with the single allowable transaction vector $\begin{bmatrix} 1 \\ -1 \end{bmatrix}$.

(iv) *Simple does not imply pure.*

To see this just look at the 2-account system with the single allowable transaction vector $\begin{bmatrix} -2 \\ 2 \end{bmatrix}$.

(v) *Inverse does not imply error-correcting.*

Consider the 3-account system with allowable transaction vectors $\mathbf{e}(1, 2)$, $\mathbf{e}(2, 3)$, $\mathbf{e}(3, 1)$. This is plainly not error-correcting. But the feasible monoid is $\mathrm{Bal}_3(\mathbb{Z})$, so the system is an inverse system.

(vi) *Elementary does not imply type-complete.*

To see this consider the 4-account system with allowable transaction vectors $\mathbf{e}(1, 2)$ and $\mathbf{e}(3, 4)$. In this system $\mathbf{e}(1, 2) + \mathbf{e}(3, 4)$

is feasible, but the vector with entries $1, -2,\ 2, -1$, which is of the
same type, is not feasible.

In conclusion we note that a variety of special classes of unbounded accounting systems have been introduced by placing restrictions on their allowable transactions. Some of the classes display features that one would expect to find in real life accounting
systems. Here we mention particularly finitely specified systems,
elementary systems and error-correcting systems. All the classes of
systems considered are natural from an algebraic standpoint and it
is at least a useful exercise to consider how far these subclasses can
be translated into practical systems, since it forces us to analyze
more precisely the true nature of such systems.

7.3. The Digraph of a Simple System

Our aim in this section is to show that the feasible digraph of
a simple unbounded accounting system has a very special form; it
consists of isolated vertices, source vertices, sink vertices and a complete digraph. Then it will be shown that a hereditary system is
determined up to equivalence by its feasible digraph and that such
systems can be completely classified.

To start things off, recall that if $\mathcal{A}$ is an unbounded accounting
system, the feasible digraph of $\mathcal{A}$ has as its vertex set the set of
accounts and there is an edge $\langle a_j, a_i \rangle$ if and only if there is a feasible
transaction vector $\mathbf{v}$ of $\mathcal{A}$ such that $v_j < 0$ and $v_i > 0$. In general
the connection between a system and its feasible digraph is quite
weak and inequivalent systems can easily have the same digraph.

The crucial property possessed by the feasible digraph of a simple
system will now be described. A digraph D is called *strongly tran-
sitive* if, whenever it contains edges $\langle d_i, d_j \rangle$ and $\langle d_k, d_\ell \rangle$ with $i \neq \ell$,
there is an edge $\langle d_i, d_\ell \rangle$ of D:

In particular this means that if D has edges $\langle d_i, d_j \rangle$ and $\langle d_j, d_\ell \rangle$
where $i \neq \ell$, then there is an edge $\langle d_i, d_\ell \rangle$.

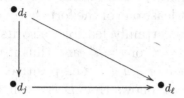

The latter property is called *transitivity* since it implies that the corresponding relation is transitive (but keep in mind that the digraph of an accounting system has no loops). These concepts are illustrated by some simple examples.

Example (7.3.1).

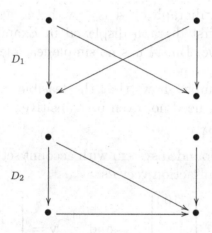

Of the digraphs above D_1 is strongly transitive, while D_2 is only transitive.

Now we come to the connection with simple systems.

(7.3.1). *Let $\mathcal{A}$ be a simple unbounded accounting system. Then the feasible digraph of $\mathcal{A}$ is strongly transitive.*

Proof

Let D be the feasible digraph of $\mathcal{A}$. To prove strong transitivity, let $\langle a_i, a_j \rangle$ and $\langle a_k, a_\ell \rangle$ be edges of D with $i \neq \ell$; it must be shown that $\langle a_i, a_\ell \rangle$ is an edge of D, and here we can assume that $i \neq k$. Now there are feasible transaction vectors $\mathbf{u}$ and $\mathbf{v}$ of $\mathcal{A}$ such that $u_i < 0$, $u_j > 0$ and $v_k < 0$, $v_\ell > 0$. Since $\mathcal{A}$ is simple, $\mathbf{u}$ is a sum of simple feasible transactions at least one of which must be of the form $a\mathbf{e}(i, i')$ where $a < 0$; also $\mathbf{v}$ is a sum of simple feasible transactions with at least one of the form $b\mathbf{e}(\ell', \ell)$ where $b < 0$. Then $a\mathbf{e}(i, i') + b\mathbf{e}(\ell', \ell)$ is certainly feasible; also its i-component is a or $a + b$, according as $i \neq \ell'$ or $i = \ell'$, and thus it is negative. Similarly its ℓ-component is $-b$ or $-a - b$, i.e., positive. Consequently D has an edge $\langle a_i, a_\ell \rangle$ and it follows that D is strongly transitive. $\square$

Notice that the converse of 7.3.1 is false. If $\mathcal{A}$ is the system with the single allowable transaction

$$\begin{bmatrix} -1 \\ 1 \\ -1 \\ 1 \end{bmatrix},$$

then the feasible digraph of $\mathcal{A}$ is – with the appropriate order of accounts – the first digraph displayed in Example 7.3.1, which is strongly transitive. But $\mathcal{A}$ has no simple feasible transactions, so it is not a simple system.

The next example shows that the feasible digraph of a finitely generated system need not even be transitive.

Example (7.3.2).
Let $\mathcal{A}$ be the unbounded system with account set $\{a_1, a_2, a_3, a_4\}$ and two allowable transaction vectors

$$\mathbf{u} = \begin{bmatrix} 100 \\ -1 \\ 1 \\ -100 \end{bmatrix} \quad \text{and} \quad \mathbf{v} = \begin{bmatrix} -100 \\ 100 \\ -1 \\ 1 \end{bmatrix}.$$

Then $\mathcal{A}$ is a finitely generated system, but in fact it is not simple. In order to see this, we first identify the feasible digraph D of $\mathcal{A}$. Now every feasible transaction is of the form

$$a\mathbf{u} + b\mathbf{v} = \begin{bmatrix} 100a - 100b \\ -a + 100b \\ a - b \\ -100a + b \end{bmatrix}, \quad a, b \geq 0.$$

Next D has edges $\langle a_1, a_2 \rangle$, $\langle a_3, a_2 \rangle$, $\langle a_1, a_4 \rangle$, $\langle a_3, a_4 \rangle$, $\langle a_2, a_1 \rangle$, $\langle a_4, a_1 \rangle$, $\langle a_2, a_3 \rangle$ and $\langle a_4, a_3 \rangle$, as one sees by inspecting the allowable transactions. Also $\langle a_4, a_2 \rangle$ is an edge, as can be seen by setting $a = 1 = b$.

On the other hand, there are no edges $\langle a_1, a_3 \rangle$ or $\langle a_3, a_1 \rangle$ since the 1- and 3-entries of $a\mathbf{u} + b\mathbf{v}$ have the same sign. Nor is there an edge $\langle a_2, a_4 \rangle$ since the inequalities $-a + 100b < 0$ and $-100a + b > 0$ are incompatible. Therefore D is the digraph below.

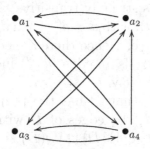

Notice that D is not even transitive since there are edges $\langle a_2, a_1 \rangle$ and $\langle a_1, a_4 \rangle$, but no edge $\langle a_2, a_4 \rangle$.

The structure of strongly transitive digraphs

Strong transitivity is such a restrictive property that it is possible to describe precisely the digraphs which have it. For this purpose some notation will be developed for an arbitrary digraph D. The vertex set $V(D)$ of D may be partitioned into four subsets,

$$V(D) = V_{is}(D) \cup V_{so}(D) \cup V_c(D) \cup V_{si}(D),$$

which are defined in the following way.

1. $V_{is}(D)$ is the set of *isolated vertices*, i.e., with in- and out-degree 0;

2. $V_{so}(D)$ is the set of *sources*, i.e., vertices with positive out-degree and zero in-degree;

3. $V_c(D)$ is the set of vertices with positive in- and out-degree;

4. $V_{si}(D)$ is the set of *sinks*, i.e., vertices with positive in-degree and zero out-degree.

Recall here that the out-degree and in-degree of a vertex v of a digraph are the respective numbers of edges with initial vertex v and final vertex v. Strongly transitive digraphs may be characterized in terms of these four sets of vertices.

(7.3.2). *A digraph D is strongly transitive if and only if there is an edge from each vertex in $V_{so}(D) \cup V_c(D)$ to each different vertex in $V_c(D) \cup V_{si}(D)$ and there are no other edges in the digraph.*

Proof

Assume first that D is strongly transitive. Let $d \in V_{so}(D) \cup V_c(D)$ and $d' \in V_c(D) \cup V_{si}(D)$, with $d \neq d'$. Then d has positive out-degree and d' positive in-degree. Hence there are edges $\langle d, d_1 \rangle$, $\langle d'_1, d' \rangle$. By strong transitivity there is an edge $\langle d, d' \rangle$ in D. Evidently all edges in D must arise in this way.

Conversely, assume that D has the property of the statement. Let $\langle d_1, d_2 \rangle$ and $\langle d_3, d_4 \rangle$ be edges of D with $d_1 \neq d_4$. Then $d_1 \in V_{so}(D) \cup V_c(D)$ and $d_4 \in V_c(D) \cup V_{si}(D)$ since d_1 has positive out-degree and d_4 has positive in-degree. Therefore there is an edge $\langle d_1, d_4 \rangle$ and thus D is strongly transitive. $\square$

Notice that the condition of 7.3.2 implies that *if D is a strongly transitive digraph, then $V_c(D)$ is a complete digraph*, i.e., there is an edge from any vertex to any other vertex in that digraph.

Example (7.3.3).

A strongly transitive digraph D with six vertices in which $V_{is}(D) = \{v_6\}$, $V_{so}(D) = \{v_1\}$, $V_c(D) = \{v_2, v_3, v_4\}$ and $V_{si}(D) = \{v_5\}$ is exhibited below:

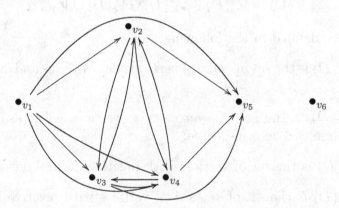

Despite the restrictive nature of the feasible digraph of a simple accounting system, it is still possible for inequivalent systems to have the same digraph.

Example (7.3.4).

Consider the elementary system $\mathcal{A}$ with account set $\{a_1, a_2, a_3, a_4\}$ which has two allowable transaction vectors $\mathbf{e}(3, 1)$ and $\mathbf{e}(4, 2)$. Here one easily sees that the feasible digraph is

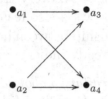

which is, of course, strongly transitive.

However, this is also the feasible digraph of the accounting system $\mathcal{A}'$ on $\{a_1, a_2, a_3, a_4\}$ whose allowable transaction vectors are $\mathbf{e}(3, 1)$, $\mathbf{e}(4, 2)$, $\mathbf{e}(4, 1)$. Now $\mathbf{e}(4, 1)$ is not feasible in $\mathcal{A}$ since it cannot be expressed in the form $a\mathbf{e}(3, 1) + b\mathbf{e}(4, 2)$ with $a, b \geq 0$. Thus $\mathcal{A}$ and $\mathcal{A}'$ have different monoids, so that they are not equivalent, yet they have the same digraph.

Suppose now that $\mathcal{A}$ is an arbitrary simple unbounded accounting system. Then by 7.3.1 the feasible digraph D of $\mathcal{A}$ has the structure indicated in 7.3.2. Let us consider what this implies about the system $\mathcal{A}$. The accounts in $V_{is}(D)$ are inactive accounts. It might be supposed that the such accounts are rare, but any account plan is likely to involve a large number of accounts since it must accommodate transactions between the firm and many outside companies. During a particular accounting period it might be the case that certain accounts are not used and these inactive accounts would show up as isolated vertices of the digraph. An account in $V_{so}(D)$ is a source of funds for the system, while an account in $V_{si}(D)$ can only receive funds; for example a bank loan that is gradually being paid off would fall into this category. All other accounts belong to the complete digraph $V_c(D)$. If a_i and a_j are distinct accounts in $V_c(D)$, then completeness implies that there is a feasible transaction of $\mathcal{A}$ that causes an outflow of value from a_i and an inflow to a_j; of course other accounts could be affected by the transaction too. So the conclusion is that in a simple unbounded system in at least part of the system arbitrary flows of value are possible between accounts.

Digraphs and hereditary accounting systems

In the final part of the chapter we turn to a particular type of simple system, namely the hereditary systems. Recall that an unbounded accounting system $\mathcal{A}$ is hereditary if, whenever $\mathbf{v}$ is a feasible transaction vector of $\mathcal{A}$, every transaction vector with type equal to or preceding $\mathrm{type}(\mathbf{v})$ is also feasible.

There is a close relationship between hereditary systems and their feasible digraphs, as is already indicated by the following result.

(7.3.3). *Let $\mathcal{A}$ be a hereditary unbounded accounting system with accounts $a_1, \ldots, a_n$. Then $\mathbf{e}(i, j)$ is a feasible transaction vector for $\mathcal{A}$ if and only if $\langle a_j, a_i \rangle$ is an edge of the feasible digraph of $\mathcal{A}$.*

Proof

If $\mathbf{e}(i, j)$ is feasible, then by definition of the feasible digraph there is an edge $\langle a_j, a_i \rangle$. Conversely, assume that $\langle a_j, a_i \rangle$ is an edge. Then there is a feasible transaction vector $\mathbf{v}$ of $\mathcal{A}$ with $v_i > 0$ and $v_j < 0$. Now $\mathrm{type}(\mathbf{e}(i, j)) \leq \mathrm{type}(\mathbf{v})$, so by the hereditary property $\mathbf{e}(i, j)$ is feasible in $\mathcal{A}$. $\square$

On the basis of this observation it is straightforward to show that a hereditary system is characterized up to equivalence by its feasible digraph.

(7.3.4). *Let $\mathcal{A}$ and $\mathcal{A}'$ be two hereditary unbounded accounting systems with account set $\{a_1, \ldots, a_n\}$. Then $\mathcal{A}$ and $\mathcal{A}'$ are equivalent if and only if they have the same feasible digraph.*

Proof

Let D and D' be the respective feasible digraphs of $\mathcal{A}$ and $\mathcal{A}'$. If $\mathcal{A}$ and $\mathcal{A}'$ are equivalent, then by 6.2.4 they have the same monoid and hence the same feasible transactions; therefore $D = D'$.

Conversely, assume that $D = D'$; thus $\langle a_j, a_i \rangle$ is an edge of D if and only if it is an edge of D'. Applying 7.3.3, we conclude that $\mathbf{e}(i, j)$ is feasible in $\mathcal{A}$ if and only if it is feasible in $\mathcal{A}'$. Since $\mathcal{A}$ and $\mathcal{A}'$ are hereditary, they are elementary, so their monoids are generated by feasible elementary vectors. It follows that $\mathrm{Mon}(\mathcal{A}) = \mathrm{Mon}(\mathcal{A}')$, so $\mathcal{A}$ and $\mathcal{A}'$ are equivalent. $\square$

This result illustrates the close relationship between a hereditary system and its feasible digraph. The next result illustrates the connection between hereditary systems and strongly transitive digraphs and, in particular, shows how to construct hereditary systems from strongly transitive graphs.

(7.3.5). *The following statements are valid:*

1. *Let $\mathcal{A}$ be a hereditary unbounded accounting system with feasible digraph D. Then D is strongly transitive and $\langle a_j, a_i \rangle$ is an edge of D if and only if $\mathbf{e}(i, j)$ is feasible in $\mathcal{A}$.*

2. *Let D be a strongly transitive digraph with finite vertex set $V(D) = \{a_1, \ldots, a_n\}$. Let $\mathcal{A}$ be the unbounded accounting system with account set $V(D)$ whose allowable transactions are the $\mathbf{e}(i, j)$ for which $\langle a_j, a_i \rangle$ is an edge of D. Then $\mathcal{A}$ is a hereditary system with feasible digraph D.*

Proof

1. This follows from 7.3.1 and 7.3.3.

2. The first step in the proof is to show that D is the feasible digraph of $\mathcal{A}$: of course the feasible digraph $\overline{D}$ of $\mathcal{A}$ contains all the edges in D. Suppose that $\langle a_i, a_j \rangle$ is an edge of $\overline{D}$. Then there exists a feasible transaction vector $\mathbf{v}$ with $v_i < 0$ and $v_j > 0$. Now $\mathbf{v}$ is a sum of allowable vectors $\mathbf{e}(r, s)$ where $\langle a_s, a_r \rangle$ is an edge in D. Hence at least one of these summands must have negative i-component. Therefore $a_i \notin V_{is}(D) \cup V_{si}(D)$, which, because of the structure of D, means that $a_i \in V_{so}(D) \cup V_c(D)$. By a similar argument $a_j \in V_c(D) \cup V_{si}(D)$. Since D is strongly transitive, 7.3.2 implies that there is an edge in D from a_i to a_j. It follows that D and $\overline{D}$ are the same digraph. It remains to prove that $\mathcal{A}$ is hereditary.

Let $\mathbf{v}$ be a feasible transaction of $\mathcal{A}$. Let U denote the set of all *non-feasible* transaction vectors $\mathbf{u}$ such that $\text{type}(\mathbf{u}) \leq \text{type}(\mathbf{v})$. Assuming that U is not empty, we can choose an element $\mathbf{u}$ of U which is minimal in the ordering of types; this is because the set of types is finite. Since $\mathbf{u} \neq \mathbf{0}$, we have $u_i > 0$ and $u_j < 0$ for some i, j. Then $v_i > 0$ and $v_j < 0$ since $\text{type}(\mathbf{u}) \leq \text{type}(\mathbf{v})$. Next $\mathbf{v}$ is a sum of allowable elementary vectors. This sum must involve vectors $\mathbf{e}(i, r)$ and $\mathbf{e}(s, j)$ for some r, s. Hence there are edges $\langle a_r, a_i \rangle$ and $\langle a_j, a_s \rangle$ in D. Because D is strongly transitive, $\langle a_j, a_i \rangle$ is an edge of D and, by definition of $\mathcal{A}$, it follows that $\mathbf{e}(i, j)$ is allowable and hence is feasible.

Now define

$$\mathbf{u}' = \begin{cases} \mathbf{u} + u_j\,\mathbf{e}(i,j) & \text{if } u_i + u_j > 0 \\ \mathbf{u} - u_i\,\mathbf{e}(i,j) & \text{if } u_i + u_j \leq 0 \end{cases}.$$

Then $u'_k = u_k$ for $k \neq i, j$; also $u'_i = u_i + u_j$ and $u'_j = 0$ if $u_i + u_j > 0$. In addition $u'_i = 0$ and $u'_j = u_i + u_j$ if $u_i + u_j \leq 0$. Clearly type$(\mathbf{u}') <$ type$(\mathbf{u})$, so $\mathbf{u}'$ is feasible by minimality of type$(\mathbf{u})$. Finally $\mathbf{u} = \mathbf{u}' + (-u_j)\,\mathbf{e}(i,j)$ if $u_i + u_j > 0$ and $\mathbf{u} = \mathbf{u}' + u_i\,\mathbf{e}(i,j)$ if $u_i + u_j \leq 0$. But $\mathbf{u}'$ and $\mathbf{e}(i,j)$ are feasible, from which it follows that $\mathbf{u}$ is feasible since $u_i > 0$, $u_j < 0$. By this contradiction the set U is empty and $\mathcal{A}$ is hereditary. □

One consequence of 7.3.5 is a method for distinguishing the hereditary systems among the elementary ones.

(7.3.6). *Let $\mathcal{A}$ be an elementary unbounded accounting system with accounts $a_1, \ldots, a_n$ and feasible digraph D. Then $\mathcal{A}$ is hereditary if and only if $\mathbf{e}(i,j)$ is feasible in $\mathcal{A}$ whenever $\langle a_j, a_i \rangle$ is an edge of D.*

Proof
The necessity of the condition being a consequence of 7.3.3, we assume it is satisfied in $\mathcal{A}$. By 7.3.1 the digraph D is strongly transitive and therefore by 7.3.5 there is a hereditary system $\overline{\mathcal{A}}$ on the same account set as $\mathcal{A}$ whose feasible digraph is D. By 7.3.3 the systems $\mathcal{A}$ and $\overline{\mathcal{A}}$ have the same feasible elementary transactions and hence the same monoid. Thus $\mathcal{A}$ and $\overline{\mathcal{A}}$ are equivalent and, since $\overline{\mathcal{A}}$ is hereditary, it follows that $\mathcal{A}$ is hereditary. □

Remark
It is *not true* that an elementary unbounded system whose feasible digraph is strongly transitive is hereditary. Indeed, in Example 7.3.4 the feasible digraph is strongly transitive, but the accounting system is not hereditary: for $\mathbf{e}(4,1)$ is not feasible, despite the existence of an edge $\langle a_1, a_4 \rangle$ in the digraph.

Counting hereditary systems

One consequence of 7.3.4 is that a hereditary system $\mathcal{A}$ is determined up to equivalence by its feasible digraph. The problem of counting non-equivalent hereditary systems is therefore reduced to that of counting the strongly transitive digraphs D with a fixed set of n vertices. Note that D is determined by the four subsets $V_{is}(D)$,

$V_{so}(D)$, $V_c(D)$, $V_{si}(D)$ of the vertex set $V(D)$. As a first step we show how to count the isomorphism types of strongly transitive digraphs. The terminology here is that two digraphs are *isomorphic* if there is a bijection between their vertex sets which preserves edge connectedness. To count the isomorphism classes we have to solve a distribution problem about placing n identical objects in four boxes subject to suitable conditions.

(7.3.7). *The number of non-isomorphic strongly transitive digraphs with n vertices is equal to*

$$1 + \frac{1}{6}(n-1)(n^2 + 7n - 6).$$

The number of connected digraphs among these is

$$\frac{1}{2}(n^2 + 3n - 6)$$

provided $n \geq 2$, and it is 1 if $n = 1$.

Proof
Let $V = \{v_1, v_2, \ldots, v_n\}$ be the set of vertices to be used. The strongly transitive digraphs on V correspond up to isomorphism to ordered partitions of V

$$V = V_{is} \cup V_{so} \cup V_c \cup V_{si}$$

where V_{is}, V_{so}, V_c, V_{si} are to be the subsets $V_{is}(D)$, $V_{so}(D)$, $V_c(D)$, $V_{si}(D)$ for the digraph D. The isomorphism type of the digraph is determined once the partition is specified. The corresponding combinatorial problem is that of placing n identical objects in four distinct boxes; in fact we have to count the solutions $(\ell_{is}, \ell_{so}, \ell_c, \ell_{si})$ in non-negative integers of the equation

$$\ell_{is} + \ell_{so} + \ell_c + \ell_{si} = n,$$

subject to two obvious restrictions:

1. $\ell_{is} \neq n - 1$;

2. if $\ell_{is} \neq n$ and $\ell_c \leq 1$, then $\ell_{so}, \ell_{si} > 0$.

If ℓ_{is} is chosen to be n, then of course $\ell_{so} = \ell_c = \ell_{si} = 0$, and there is just one possible solution. Assume therefore that $\ell_{is} \leq n - 2$ and suppose ℓ_{is} has been chosen to be j where $0 \le j \le n - 2$. Then we have to solve

$$\ell_{so} + \ell_c + \ell_{si} = n - j.$$

If $\ell_c = 0$, there are $n - j - 1$ choices for ℓ_{so} and ℓ_{si} since by the second condition neither can be zero. If $\ell_c = 1$, there are $n - j - 2$ choices for the same reason.

Next assume that $\ell_c \ge 2$. The problem is now that of placing $n - j$ identical objects in three boxes with at least two objects in the box corresponding to ℓ_c. By a well-known combinatorial formula – see [2] – the number of ways to do this is

$$\binom{(n - j - 2) + 3 - 1}{n - j - 2} = \binom{n - j}{n - j - 2} = \binom{n - j}{2}.$$

Adding up the numbers of distributions for $j = 0, 1, \ldots, n - 2$ and remembering the single distribution for $j = \ell_{is} = n$, we obtain as the number of distributions

$$1 + \sum_{j=0}^{n-2} \left((n - j - 1) + (n - j - 2) + \binom{n - j}{2} \right),$$

which equals

$$1 + \frac{1}{2}n(n - 1) + \frac{1}{2}(n - 1)(n - 2) + \binom{n + 1}{3},$$

since $\binom{2}{2} + \binom{3}{2} + \cdots + \binom{n}{2} = \binom{n+1}{3}$. After simplification, the number of distributions is found to be $1 + \frac{1}{6}(n - 1)(n^2 + 7n - 6)$.

Next we count the connected digraphs: for these V_{is} is empty, i.e., $\ell_{is} = 0$. If $n = 1$, the number of these is obviously 1, so let $n \ge 2$. Now $\ell_{is} = 0$, so we have to solve $\ell_{so} + \ell_c + \ell_{si} = n$. If $\ell_c = 0$, these are $n - 1$ solutions and if $\ell_c = 1$, there are $n - 2$. If $\ell_c \ge 2$, then the number of solutions is

$$\binom{(n - 2) + 3 - 1}{n - 2} = \binom{n}{2}$$

by the distribution argument above. Hence the number of indecomposable digraphs is

$$(n - 1) + (n - 2) + \binom{n}{2} = \frac{1}{2}(n^2 + 3n - 6). \qquad \square$$

Example (7.3.5).

By 7.3.7 there are nine isomorphism types of strongly transitive digraphs with three vertices and six of these are connected. These digraphs are displayed in the following list.

(i)

$$|V_{is}(D)| = 0 \quad |V_{so}(D)| = 0 \quad |V_c(D)| = 3 \quad |V_{si}(D)| = 0.$$

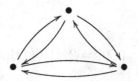

(ii)

$$|V_{is}(D)| = 0 \quad |V_{so}(D)| = 0 \quad |V_c(D)| = 2 \quad |V_{si}(D)| = 1.$$

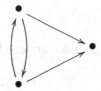

(iii)

$$|V_{is}(D)| = 0 \quad |V_{so}(D)| = 1 \quad |V_c(D)| = 2 \quad |V_{si}(D)| = 0.$$

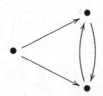

(iv)

$$|V_{is}(D)| = 0 \quad |V_{so}(D)| = 1 \quad |V_c(D)| = 1 \quad |V_{si}(D)| = 1.$$

(v)

$$|V_{is}(D)| = 0 \quad |V_{so}(D)| = 1 \quad |V_c(D)| = 0 \quad |V_{si}(D)| = 2.$$

(vi)

$$|V_{is}(D)| = 0 \quad |V_{so}(D)| = 2 \quad |V_c(D)| = 0 \quad |V_{si}(D)| = 1.$$

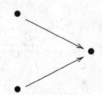

The remaining digraphs are disconnected.

(vii)

$$|V_{is}(D)| = 1 \quad |V_{so}(D)| = 0 \quad |V_c(D)| = 2 \quad |V_{si}(D)| = 0.$$

(viii)

$$|V_{is}(D)| = 1 \quad |V_{so}(D)| = 1 \quad |V_c(D)| = 0 \quad |V_{si}(D)| = 1.$$

(ix)

$$|V_{is}(D)| = 3 \quad |V_{so}(D)| = 0 \quad |V_c(D)| = 0 \quad |V_{si}(D)| = 0.$$

$$\bullet \qquad \bullet \qquad \bullet$$

Of course these digraphs are not yet labeled, which means that the accounting systems are not fully specified. For each type of digraph we have to label the vertices by the accounts a_1, a_2, a_3 in all possible ways. This is easy to do in this instance since the possibilities are very limited. A quick check of the digraphs reveals that the numbers of labeled digraphs of each respective type are $1, 3, 3, 6, 3, 3, 3, 6, 1$, giving a total of 29 labeled digraphs. Hence, up to equivalence there are 29 hereditary accounting systems with three accounts. The count of labeled connected digraphs is $1, 3, 3, 6, 3, 3$, giving 19 in all. Thus there are 19 indecomposable hereditary accounting systems with three accounts.

It is natural to enquire if an explicit formula for the number of equivalence classes of hereditary accounting systems can be found. The answer is affirmative, but it involves a more complex distribution problem in which the distributed objects, the accounts, are all different. There is standard procedure in combinatorics for solving such problems which calls for the use of *exponential generating functions*. These are formal power series of the form

$$1 + \frac{c_1 x}{1!} + \frac{c_2 x^2}{2!} + \cdots + \frac{c_n x^n}{n!} + \cdots$$

in which the coefficient c_n of $\frac{x^n}{n!}$ is the number of distributions sought in the problem: here the $n!$ is inserted to allow for permutations of the n objects being distributed. For a detailed account of exponential generating functions see [2].

The definitive result is:

(7.3.8). *The number of equivalence classes of unbounded hereditary accounting systems with a fixed set of n accounts is*

$$4^n - (n + 2)2^n + n + 2.$$

The number of indecomposable systems among these is

$$3^n - 2n - 2,$$

provided that $n \geq 2$. When $n = 1$, the number is 1.

Proof

We use the notation of the proof of 7.3.7: here too it is necessary to treat separately the cases $\ell_c = 0, 1$ and $\ell_c \geq 2$.

Suppose first that $\ell_c = 0$. The problem is to place n different objects in three boxes with two of the boxes non-empty. The generating function for this is

$$e^x(e^x - 1)^2 = e^{3x} - 2e^{2x} + e^x,$$

where, of course, e^x is the exponential function

$$1 + \frac{x}{1!} + \frac{x^2}{2!} + \cdots + \frac{x^n}{n!} + \cdots .$$

The coefficient of $\frac{x^n}{n!}$ in the generating function is clearly $3^n - 2^{n+1} + 1$. Next suppose that $\ell_c = 1$. Now only $n-1$ objects are to be placed in the boxes, so the number is $3^{n-1} - 2^n + 1$, but this must be multiplied by n since there are n choices for the single object to go in box V_c.

Now suppose that $\ell_c \geq 2$. In this case there is no restriction on $\ell_{is}, \ell_{si}, \ell_{so}$, so that the generating function is

$$(e^x - 1 - x)(e^x)^3 = e^{4x} - e^{3x} - xe^{3x}$$

and the coefficient of $\frac{x^n}{n!}$ is $4^n - 3^n - n3^{n-1}$. If these numbers are added up and the single case $\ell_{is} = n$ counted, we obtain after cancelation $4^n - (n+2)2^n + n + 2$, as claimed.

Turning to the connected case, we have $\ell_{is} = 0$. Again it is necessary to distinguish the values of ℓ_c. When ℓ_c equals 0 or 1, the generating function is $(e^x - 1)^2$ and the respective coefficients of $\frac{x^n}{n!}$ and $\frac{x^{n-1}}{(n-1)!}$ are $2^n - 2$ and $2^{n-1} - 2$, but the second of these must be multiplied by n. Let $\ell_c \geq 2$; then the generating function is

$$(e^x - 1 - x)(e^x)^2 = e^{3x} - e^{2x} - xe^{2x},$$

so the coefficient of $\frac{x^n}{n!}$ is $3^n - n2^{n-1} - 2^n$. Addition of these numbers yields $3^n - 2n - 2$, provided $n \geq 2$: when $n = 1$, the answer is obviously 1. $\square$

For example, if we set $n = 3$ in 7.3.8, we obtain 29 and 19 as the numbers of systems and indecomposable systems respectively, confirming what was found in Example 7.3.5.

Having seen that hereditary accounting systems have a nice description in terms of their feasible digraphs, one may wonder whether a real life system could be of this type. Hereditary systems have the property that, given active accounts a_i and a_j which are neither sources or sinks, there is a feasible transaction vector $\mathbf{e}(i, j)$; thus one can by a suitable sequence of allowable transactions arrange for a transfer of value from a_j to a_i, without in the end changing the balances of other accounts. Thus a hereditary system has a high degree of fluidity built into its structure, a feature which could be useful in real life systems, but which might require strong controls to reduce the risk of inappropriate operations being applied to the system.

Chapter Eight

Algorithms

8.1. Decision Problems for Accounting Systems

In the previous chapter a large number of special types of accounting system were introduced by restricting the allowable transactions in various ways. The motivation behind this study was to discover which types of system are closest to reality. Continuing this line of enquiry, we observe that one natural test of the practicality of a proposed model is whether it is possible, in principle at least, to perform routine operations on the system, including checks and security procedures, by machine implementation. As a first step let us identify some of the procedures one would expect to be able to implement for an accounting system.

1. Decide whether a given transaction is allowable.

2. Decide whether a given balance vector is allowable.

3. Decide whether a given transaction is feasible.

4. Decide whether a final balance vector could actually have occurred by correctly applying a sequence of allowable transactions to a given initial balance vector.

5. Decide whether two accounting systems on the same account set are equivalent, i.e., if they have the same feasible transactions and hence the same monoid.

6. Decide whether a given accounting system is of a specific type such as those described in Chapter 7.

First of all, some explanation is called for regarding what is meant by a "decision problem" like those listed here. In general it is possible to *decide* whether a statement is true or false if there is an *algorithm* which will give a definite "yes" or "no" to the question. Here the term "algorithm" is used in the classical sense: there is an algorithm to produce a set of data if the data can be obtained from the outputs of a finite number of Turing machines. Here a *Turing machine* can be thought of as an abstract model of a computer, related to, but more powerful than, a finite state automaton.

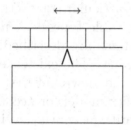

Specifically, a Turing machine consists of a "head" and a "tape" divided into squares each of which has a symbol written on it. The head is able to read symbols on the squares. At any instant the machine is in one of a finite number of states and the set of symbols is finite. It scans a square on the tape and reads the symbol on it; as a result it goes to new state and writes another symbol on the square. The head then moves either left or right by one square and repeats the procedure. Turing machines are fundamental in the modern theory of computability: see [3], [4] or [5] for details.

Notice that there is no requirement restricting the number of steps involved in an algorithm; it is sufficient to know that it will halt after finitely many steps. Such algorithms might therefore require large amounts of computing power – and indeed some of those described below surely do. However, with the enormous increase in computing power in recent years this is not necessarily a serious objection.

Several of the decision problems listed above lead directly to questions about finite systems of linear equations over the integers. The problem is usually to determine if the system has a solution in non-negative integers. Fortunately this is an area of applied algebra which has been extensively developed. Such problems are special types of linear programming problems called *integer programs*. Some highly effective algorithms for solving such problems are available and may be applied in our situation.

In Section 8.2 we consider some wide classes of accounting systems for which it is possible to decide if a given transaction vector or balance vector is allowable. At the same time examples are given which show that there are limits to what is computable in general accounting systems.

Section 8.3 is concerned with accounting systems which are finitely specifiable, i.e., equivalent to finitely specified systems (in which there are finitely many allowable particular transactions, in addition to all transactions of certain allowable types). A number of algorithms for finitely specifiable systems are described, most of them being based on integer programming. The existence of these practical algorithms lends support to the viewpoint that finitely specifiable systems should be our premier model for accounting systems. Finally, Section 8.4 describes algorithms which can, in favorable circumstances, decide if an accounting system is inverse, elementary, or hereditary.

8.2. Recursive Accounting Systems

We begin with a review of some basic terminology from recursion theory: as references for this we cite [3] and [4]. Let

$$\mathbb{P}$$

denote the set of positive integers. A subset S of $\mathbb{P}$ is called *recursively enumerable* if its elements are the output of some Turing machine. Equivalently, one could say that S is the set of values of a partial recursive function. In practice it is usually convenient to think of the elements of S as being "enumerated" by a Turing machine in the form of a sequence $s_1, s_2, \ldots$.

Next a subset S of $\mathbb{P}$ is said to be *recursive* if both S and its complement $\mathbb{P} \backslash S$ are recursively enumerable. Thus there are Turing machines that can enumerate the elements of S and also the elements of its complement. Another way to express this is to say that S is recursive if and only if there is an algorithm which, when a positive integer n is given, decides whether or not n belongs to S. The last statement is often referred to as asserting that *the membership problem* for the subset S is solvable.

It is easy to see that there are subsets of $\mathbb{P}$ which are not recursively enumerable. For the number of Turing machines (or algorithms) is surely countable since each one is determined by a finite list of rules. On the other hand, there are uncountably many subsets of $\mathbb{P}$, so some of them – in fact uncountably many – must fail be recursively enumerable. It is much harder to show that there exist recursively enumerable subsets of $\mathbb{P}$ which are not recursive: for such a subset S it is possible to enumerate the elements of S (as the output of Turing machines), but not those of its complement $\mathbb{P}\backslash S$. Thus there is no algorithm which can decide if an integer is *not* in S. For this fundamental result see [3] or [4].

So far the terms recursively enumerable and recursive have been applied to subsets of $\mathbb{P}$, but they can equally well be applied to the subsets of any countably infinite set U since the elements of U may be labeled by positive integers $\{u_i \mid i \in \mathbb{P}\}$. This usage is important since it will permit us to speak of recursive and recursively enumerable subsets of $\mathrm{Bal}_n(\mathbb{Z})$.

These concepts will be used to introduce two wide classes of accounting systems that may be thought of as the most general systems to which algorithms can be applied in any useful way.

Definitions

Let $\mathcal{A} = (A \mid T \mid B)$ be an accounting system over $\mathbb{Z}$ with n accounts. If T and B are recursively enumerable subsets of $\mathrm{Bal}_n(\mathbb{Z})$, then $\mathcal{A}$ is called a *recursively enumerable accounting system*. If T and B are recursive subsets of $\mathrm{Bal}_n(\mathbb{Z})$, then $\mathcal{A}$ is said to be a *recursive accounting system*. Clearly a recursive system is recursively enumerable, but the converse is false.

Note on the domain of account values

The preceding definitions have been formulated for accounts over $\mathbb{Z}$, but they can be stated so as to apply to an accounting system over an ordered domain R which is *computable*. Roughly speaking, this means that R is countable, the ring operations of addition, multiplication and the formation of negatives in R are computable by Turing machines, and the identity problem for R is solvable: the last statement means that given elements r_1, r_2 of R, it is possible to decide if $r_1 \neq r_2$. It is evident that $\mathbb{Z}$ is a computable ring in this sense. It is known that every finitely generated commutative ring is computable, so there are many possibilities for R which might serve as an appropriate domain for account values. However, in the interest of simplicity we will assume throughout this chapter that *all accounting systems are over* $\mathbb{Z}$.

Algorithms for recursive systems

The most basic algorithmic properties that one would require an accounting system to have are the ability to list allowable transactions and balance vectors, to decide if a given transaction and the resulting balance vector are allowable, and, if so, to apply the transaction and compute the new balance vector. Recursively enumerable systems and recursive systems are characterized by these properties.

(8.2.1). *Let $\mathcal{A}$ be an accounting system.*

1. *There are algorithms which can enumerate the allowable transactions and the allowable balances of $\mathcal{A}$ if and only if $\mathcal{A}$ is a recursively enumerable system.*

2. *There are algorithms which can decide if a given transaction is allowable and if a given balance vector is allowable if and only if $\mathcal{A}$ is a recursive system.*

This result is an immediate consequence of the definitions given above. Recursive systems have the crucial ability to accept or reject a transaction, and in case of acceptance to compute the new balance.

(8.2.2). *Let $\mathcal{A}$ be a recursive accounting system with n accounts. Then there is an algorithm which, when given vectors $\mathbf{b}(0)$ and $\mathbf{v}$ in $\mathrm{Bal}_n(\mathbb{Z})$, with $\mathbf{b}(0)$ the current balance of $\mathcal{A}$, either rejects the transaction $\mathbf{v}$ or else accepts it and records $\mathbf{b}(0) + \mathbf{v}$ as the new balance vector.*

Proof
Let $\mathcal{A} = (A|\ T|\ B)$; thus T and B are recursive subsets of $\mathrm{Bal}_n(\mathbb{Z})$. The algorithm decides whether $\mathbf{v}$ belongs to T and, if this is true, it then decides if $\mathbf{b}(0) + \mathbf{v}$ belongs to B; there are algorithms to perform these actions since T and B are recursive sets. If the answer is positive in both cases, then $\mathbf{b}(1) = \mathbf{b}(0) + \mathbf{v}$ is the new balance vector. Otherwise the balance vector remains $\mathbf{b}(0)$. $\square$

Notice that 8.2.2 mirrors the operation of the automaton of 6.2: the additional information provided here is that the system can be operated by using Turing machines. It is important to realize that there are accounting systems which are not recursively enumerable, and also recursively enumerable systems which are not recursive.

Example (8.2.1).
Let S_1 be a non-recursively enumerable subset of $\mathbb{P}$ – recall that there are uncountably many of these. Let $\mathcal{A}_1 = (A| \ T_1| \ B_1)$ be the 2-account system for which

$$T_1 = B_1 = \left\{ \begin{bmatrix} a \\ -a \end{bmatrix} \ \middle| \ a \in S_1 \right\}.$$

The system $\mathcal{A}_1$ is not recursively enumerable: for if it were possible to enumerate the elements of T_1 by means of a Turing machine, the same would be true of the elements of S_1.

Example (8.2.2).
Let S_2 be a recursively enumerable, but non-recursive subset of $\mathbb{P}$. Define an accounting system $\mathcal{A}_2 = (A| \ T_2| \ B_2)$ on two accounts by

$$T_2 = B_2 = \left\{ \begin{bmatrix} a \\ -a \end{bmatrix} \ \middle| \ a \in S_2 \right\}.$$

Then $\mathcal{A}_2$ is recursively enumerable since T_2 is, but it is not recursive since T_2 is not be recursive.

Of course the accounting systems appearing in these examples are of purely theoretical interest, but their inclusion here serves to demarcate the limits of computability in accounting systems.

We move on to consider which types of system introduced in Chapter 7 have good algorithmic properties. It is reassuring that many types of finitely specifiable system are recursive.

(8.2.3). *A finitely specified system is recursive if and only if it has a recursive set of allowable balance vectors.*

Proof
Let $\mathcal{A} = (A| \ T_0, T_1| \ B)$ where T_0 is the set of allowable transaction types, T_1 is a *finite* set of allowable vectors and B is the set of all allowable balance vectors. Suppose that B is recursive. Assume that $\mathcal{A}$ has n accounts and that $\mathbf{v} \in \mathrm{Bal}_n(\mathbb{Z})$ is given. The algorithm to decide if $\mathbf{v}$ is allowable in $\mathcal{A}$ proceeds as follows: first it determines if $\mathrm{type}(\mathbf{v})$ belongs to T_0 and if not, whether $\mathbf{v}$ belongs to T_1. Note that

T_0 and T_1 are finite sets, so this is certainly possible. This means that we can decide whether $\mathbf{v}$ is an allowable transaction vector. Next let $\mathbf{b} \in \text{Bal}_n(\mathbb{Z})$ be given; since B is recursive, there is an algorithm to decide if $\mathbf{b} \in B$, i.e., whether $\mathbf{b}$ is an allowable balance vector for $\mathcal{A}$. Therefore $\mathcal{A}$ is a recursive system. The converse is clearly true. $\square$

For example, a finitely specified system is recursive if either (i) it is unbounded, i.e., all balance vectors are allowable, or (ii) it is absolutely bounded, so there are just finitely many allowable balance vectors.

A natural extension of these problems is to decide whether a given balance vector is feasible for a system, i.e., whether it is obtainable by a sequence of allowable transactions. This will be considered in the following section.

8.3. The Balance Verification Problem

An important problem for accounting systems is the construction of an algorithm which can verify balances. More precisely, suppose that an accounting system $\mathcal{A}$ had an initial balance vector $\mathbf{b}^{(i)}$ and that the final balance vector at the end of some period of time is $\mathbf{b}^{(f)}$. The critical question is whether $\mathbf{b}^{(f)}$ could really have been obtained from $\mathbf{b}^{(i)}$ by applying a finite sequence of allowable transactions of $\mathcal{A}$, while satisfying the balance restrictions of $\mathcal{A}$. This will be called the *balance verification problem* for $\mathcal{A}$. If the balance verification problem can be solved for the system $\mathcal{A}$, then the associated algorithm will provide a useful safeguard against misuse or malfunction of the system.

In the interest of simplicity we assume that $\mathcal{A}$ is an unbounded accounting system with n accounts: we will have something to say about bounded systems later. Suppose that $\mathbf{b}^{(i)}$ and $\mathbf{b}^{(f)}$ are the respective initial and final balance vectors of $\mathcal{A}$ over some period. For $\mathbf{b}^{(f)}$ to be a legitimate final balance vector, there must exist a sequence of allowable transaction vectors $\mathbf{v}(1), \mathbf{v}(2), \ldots, \mathbf{v}(m)$ such that, when the corresponding transactions are applied in sequence to $\mathbf{b}^{(i)}$, the final balance vector $\mathbf{b}^{(f)}$ is obtained. Thus

$$f_{\mathbf{v}(m)} \circ f_{\mathbf{v}(m-1)} \circ \cdots \circ f_{\mathbf{v}(1)}(\mathbf{b}^{(i)}) = \mathbf{b}^{(f)},$$

where the function $f_{\mathbf{v}}$ is given by the equation $f_{\mathbf{v}}(\mathbf{b}) = \mathbf{b} + \mathbf{v}$ since there are no balance restrictions for $\mathcal{A}$. This is equivalent to requiring that $\mathbf{b}^{(f)} = \mathbf{b}^{(i)} + \mathbf{v}(1) + \cdots + \mathbf{v}(m)$, that is

$$\mathbf{b}^{(f)} - \mathbf{b}^{(i)} = \mathbf{v}(1) + \cdots + \mathbf{v}(m),$$

which simply states that $\mathbf{b}^{(f)} - \mathbf{b}^{(i)}$ must belong to $\mathrm{Mon}(\mathcal{A})$. Conversely, if this conclusion is true, then $\mathbf{b}^{(f)} - \mathbf{b}^{(i)}$ is a sum of allowable vectors and we see by reversing the argument above that $\mathbf{b}^{(f)}$ is a legitimate final balance vector.

What the preceding discussion shows is that the balance verification problem for an unbounded system $\mathcal{A}$ is equivalent to the problem of deciding whether a given element of $\mathrm{Bal}_n(\mathbb{Z})$ belongs to the submonoid $\mathrm{Mon}(\mathcal{A})$. This is *the membership problem* for $\mathrm{Mon}(\mathcal{A})$. Another formulation of the problem is as *the feasibility problem* for $\mathcal{A}$, which is to decide if a given transaction is feasible for $\mathcal{A}$. Therefore we have the following result.

(8.3.1). *For an unbounded accounting system $\mathcal{A}$ the following statements are equivalent:*

1. *the balance verification problem is solvable for $\mathcal{A}$;*

2. *the membership problem for $\mathrm{Mon}(\mathcal{A})$ is solvable;*

3. *the feasibility problem for $\mathcal{A}$ is solvable.*

A noteworthy consequence of this result is:

(8.3.2). *Let $\mathcal{A}$ and $\mathcal{A}'$ be two equivalent unbounded accounting systems. If the balance verification problem is solvable for $\mathcal{A}$, then it is solvable for $\mathcal{A}'$.*

The reason for this is that by definition $\mathrm{Mon}(\mathcal{A}) = \mathrm{Mon}(\mathcal{A}')$ and the second version of the balance verification problem yields the result.

Verifying balances in finitely specifiable systems

It is an important property of finitely specifiable accounting systems that the balance verification problem is always solvable. What is more, the solution involves an efficient algorithm.

(8.3.3). *Let $\mathcal{A}$ be a finitely specifiable, unbounded accounting system. Then the balance verification problem is solvable for $\mathcal{A}$.*

Proof

In the first place we observe that $\mathcal{A}$ is equivalent to a finitely *specified* system $\mathcal{A}'$. Then, on the basis of 8.3.2, we see that it suffices to prove the result for $\mathcal{A}'$, so there is no loss in assuming that $\mathcal{A}$ is a finitely specified system, say $\mathcal{A} = (A|\ T_0, T_1)$, where as usual T_0 is a set of allowable transaction types and T_1 is a *finite* set of allowable transaction vectors. It is assumed that we have explicit knowledge of the sets T_0 and T_1, which of course constitutes a finite amount of data.

According to 8.3.1 the balance verification problem for $\mathcal{A}$ is equivalent to the membership problem for the submonoid $\mathrm{Mon}(\mathcal{A})$. Therefore the problem is to find an algorithm which, when a vector $\mathbf{v} \in \mathrm{Bal}_n(\mathbb{Z})$ is given, decides if $\mathbf{v} \in M = \mathrm{Mon}(\mathcal{A})$: here n is the number of accounts in $\mathcal{A}$. Let

$$T_0 = \{\mathbf{t}(1), \ldots, \mathbf{t}(r)\} \quad \text{and} \quad T_1 = \{\mathbf{v}(1), \ldots, \mathbf{v}(s)\}$$

where $\mathbf{t}(i)$ is a transaction type, (i.e., a column vector with entries 0, $+$ or $-$), and $\mathbf{v}(1), \ldots, \mathbf{v}(s)$ are given vectors in $\mathrm{Bal}_n(\mathbb{Z})$. Of course these vectors are assumed to be known. Then $\mathbf{v} \in M$ if and only if there is an expression

$$\mathbf{v} = \mathbf{w}(1) + \cdots + \mathbf{w}(r) + \ell_1\mathbf{v}_1(1) + \cdots + \ell_s\mathbf{v}(s),$$

where $\mathbf{w}(i) \in \mathrm{Bal}_n(\mathbb{Z})$ is of type $\mathbf{t}(i)$ and the ℓ_j are non-negative integers. Thus the problem is to decide whether or not such an expression exists.

The sign of an entry of $\mathbf{w}(i)$ is determined by the corresponding entry of its type $\mathbf{t}(i)$. Let the positive entries of $\mathbf{w}(i)$ be $1 + x_{ij}$, $j = 1, 2, \ldots, p_i$, and the negative entries $-1 - y_{ij}$, $j = 1, 2, \ldots, q_i$, where $x_{ij}, y_{ij} \geq 0$, all other entries of $\mathbf{w}(i)$ being 0. Equating corresponding entries on each side of the equation for $\mathbf{v}$, we obtain a system of n linear equations over $\mathbb{Z}$ for the $s + \sum_{i=1}^r (p_i + q_i)$ unknowns x_{ij}, y_{ij}, ℓ_k. Also, it is necessary to adjoin the equations

$$\sum_{j=1}^{p_i}(1 + x_{ij}) - \sum_{j=1}^{q_i}(1 + y_{ij}) = 0, \quad i = 1, \ldots, r,$$

these being the conditions for the $\mathbf{w}(i)$ to be balance vectors. For there to be an expression for $\mathbf{v}$ of the type just considered, it must be possible to find a solution of the linear system of $n+r$ equations in *non-negative integers* x_{ij}, y_{ij}, ℓ_k.

The foregoing argument shows that in order to solve our problem we need a way of determining whether a linear system of equations over $\mathbb{Z}$ has a non-negative integer solution. This is an *integer programming* problem. There are several efficient algorithms available for solving integer programs, of which the best known is *Gomery's fractional algorithm*. It is essentially a refinement of the well known simplex algorithm designed to eliminate fractional solutions. The proof of the theorem can therefore be completed by invoking the existence of such an algorithm.   □

The method of proof of 8.3.3 will now be illustrated with a numerical example: to follow all the details a knowledge of integer programming is necessary – see for example [6].

Example (8.3.1).
An unbounded accounting system $\mathcal{A}$ with three accounts has one allowable transaction type and one explicit allowable transaction,

$$\begin{bmatrix} - \\ + \\ + \end{bmatrix} \quad \text{and} \quad \begin{bmatrix} -100 \\ -100 \\ 200 \end{bmatrix}.$$

The balance vectors of $\mathcal{A}$ at the beginning and end of an accounting period are recorded as

$$\begin{bmatrix} 100 \\ -100 \\ 0 \end{bmatrix} \quad \text{and} \quad \begin{bmatrix} -300 \\ -200 \\ 500 \end{bmatrix},$$

respectively. The question is whether the latter is in fact a possible final balance vector for $\mathcal{A}$.

The problem here is to decide if the vector

$$\mathbf{v} = \begin{bmatrix} -300 \\ -200 \\ 500 \end{bmatrix} - \begin{bmatrix} 100 \\ -100 \\ 0 \end{bmatrix} = \begin{bmatrix} -400 \\ -100 \\ 500 \end{bmatrix}$$

is feasible in $\mathcal{A}$: for then there will be a sequence of allowable transactions which transform the initial balance vector into the final one.

Now $\mathbf{v}$ is feasible if and only if there is an expression

$$\mathbf{v} = \ell_1 \begin{bmatrix} -100 \\ -100 \\ 200 \end{bmatrix} + \begin{bmatrix} -1 - y_{11} \\ 1 + x_{11} \\ 1 + x_{12} \end{bmatrix}$$

where $\ell_1, x_{11}, x_{12}, y_{11}$ are non-negative integers. The conditions for this to hold are that

$$\begin{cases} -100\,\ell_1 - 1 - y_{11} & = -400 \\ -100\,\ell_1 + 1 + x_{11} & = -100 \, . \\ 200\,\ell_1 + 1 + x_{12} & = 500 \end{cases}$$

Notice that addition of these equations yields

$$(-1 - y_{11}) + (1 + x_{11}) + (1 + x_{12}) = 0,$$

which is the condition for the second vector in the expression for $\mathbf{v}$ to be a balance vector. Therefore all we need to do in this case is determine if the linear system

$$\begin{cases} 100\,\ell_1 + y_{11} & = 399 \\ 100\,\ell_1 - x_{11} & = 101 \, . \\ 200\,\ell_1 + x_{12} & = 499 \end{cases}$$

has a solution for $\ell_1, y_{11}, x_{11}, x_{12}$ in non-zero integers. One of the integer programming algorithms can be applied to show that there are non-negative integral solutions of this linear system: in fact

$$\ell_1 = 2, \ x_{11} = 99, \ x_{12} = 99, \ y_{11} = 199$$

is a solution. Thus

$$\mathbf{v} = 2 \begin{bmatrix} -100 \\ -100 \\ 200 \end{bmatrix} + \begin{bmatrix} -200 \\ 100 \\ 100 \end{bmatrix} ,$$

so that $\mathbf{v} \in \mathrm{Mon}(\mathcal{A})$ and $\mathbf{v}$ is feasible. The conclusion is therefore that $\begin{bmatrix} -300 \\ -200 \\ 500 \end{bmatrix}$ is indeed a possible final balance vector for the system: transactions which produce this balance are (in any order)

$$\begin{bmatrix} -200 \\ -200 \\ 400 \end{bmatrix} \text{ and } \begin{bmatrix} -200 \\ 100 \\ 100 \end{bmatrix} .$$

There is a significant application of 8.3.3 to the equivalence problem for finitely generated systems.

(8.3.4). *There is an algorithm which, when two finitely generated, unbounded accounting systems $\mathcal{A}$ and $\mathcal{A}'$ with the same account set are given, decides if they are equivalent.*

Proof

By hypothesis both $\mathrm{Mon}(\mathcal{A})$ and $\mathrm{Mon}(\mathcal{A}')$ can be generated by finitely many allowable balance vectors, say by $\mathbf{v}(1), \ldots, \mathbf{v}(m)$ and $\mathbf{v}(1)', \ldots, \mathbf{v}(m')'$ respectively. It is assumed that these vectors are known explicitly. Now $\mathcal{A}$ and $\mathcal{A}'$ are equivalent if and only if $\mathrm{Mon}(\mathcal{A})$ and $\mathrm{Mon}(\mathcal{A}')$ are equal, i.e., $\mathrm{Mon}(\mathcal{A}) \subseteq \mathrm{Mon}(\mathcal{A}')$ and $\mathrm{Mon}(\mathcal{A}') \subseteq \mathrm{Mon}(\mathcal{A})$. Hence $\mathcal{A}$ and $\mathcal{A}'$ are equivalent precisely when each $\mathbf{v}(i)$ belongs to $\mathrm{Mon}(\mathcal{A}')$ and each $\mathbf{v}(j)'$ belongs to $\mathrm{Mon}(\mathcal{A})$. This is decidable by 8.3.3. □

Verifying balances with balance restrictions

It is a more difficult problem to construct an algorithm which can check the validity of a final balance vector when the accounting system has balance restrictions. The reason is that, in addition to finding a sequence of allowable transactions leading from the initial balance vector to the final one, it is necessary to verify that all the intermediate balances that appear are allowable. As a consequence the order in which the transactions are applied is significant.

In order to solve the balance verification problem it may be necessary to examine *all* sequences of allowable transactions that lead from the initial to the final balance vector. In the case of a finitely specified system such an examination may be impossible if there are infinitely many allowable balance vectors. On the other hand, if the system has only finitely many allowable balance vectors, then it is impossible to apply to the system *all* transaction vectors of any one type, since these are infinite in number. Thus we might as well exclude allowable types from the system, in which case there are only finitely many allowable transactions and the system is finitely generated. For this reason attention is directed at finitely generated systems.

(8.3.5). *Let $\mathcal{A} = (A \mid T \mid B)$ be an accounting system with T and B both finite. Then the balance verification problem is solvable for $\mathcal{A}$.*

Proof
Let n be the number of accounts in $\mathcal{A}$. Suppose we are given vectors $\mathbf{b}^{(i)}$, $\mathbf{b}^{(f)}$, representing the initial and final balance vectors over some period; these of course should belong to B. To decide if $\mathbf{b}^{(f)}$ is a legitimate balance vector for $\mathcal{A}$, we have to consider all sequences of allowable transactions $\mathbf{v}(1), \mathbf{v}(2), \ldots, \mathbf{v}(k)$ such that

$$f_{\mathbf{v}(k)} \circ f_{\mathbf{v}(k-1)} \circ \cdots \circ f_{\mathbf{v}(1)}(\mathbf{b}^{(i)}) = \mathbf{b}^{(f)},$$

where $f_{\mathbf{v}}(\mathbf{b}) = \mathbf{b} + \mathbf{v}$ if $\mathbf{v} \in T$ and $\mathbf{b} + \mathbf{v} \in B$. The sequence produces intermediate balance vectors $\mathbf{b}(0) = \mathbf{b}^{(i)}$, $\mathbf{b}(1), \ldots, \mathbf{b}(k) = \mathbf{b}^{(f)}$ where

$$\mathbf{b}(j) = f_{\mathbf{v}(j)}(\mathbf{b}(j-1)), \quad j = 1, 2, \ldots, k.$$

For $\mathbf{b}^{(f)}$ to be an acceptable final balance there must be a sequence $\{\mathbf{v}(j)\}$ such that all the $\mathbf{b}(j)$ belong to B.

Now if such a sequence of allowable transactions exists, there is one of shortest length, say $\mathbf{v}(1), \ldots, \mathbf{v}(k)$, with an associated sequence of balance vectors $\mathbf{b}(1), \ldots, \mathbf{b}(k)$: we can assume that $k > 0$ here. Suppose that $\mathbf{b}(j) = \mathbf{b}(j')$ where $j < j'$. Then the transactions $\mathbf{v}(j+1), \ldots, \mathbf{v}(j')$ can be deleted from the sequence, leaving a sequence of shorter length which still leads from $\mathbf{b}^{(i)}$ to $\mathbf{b}^{(f)}$. By this contradiction the balance vectors $\mathbf{b}(0), \mathbf{b}(1), \ldots, \mathbf{b}(k)$ are all different and as a result we can derive the inequality $k + 1 \leq |B|$. Therefore the number ℓ of sequences that need to be examined satisfies

$$\ell \leq |T|^k \leq |T|^{|B|-1}.$$

Next let $\mathbf{v}(1), \ldots, \mathbf{v}(k)$ be one of the shortest sequences of allowable transactions to be screened and let $\mathbf{b}(j)$, $j = 0, 1, \ldots, k$, be the associated intermediate balance vectors. Each of the balance vectors $\mathbf{b}(j)$ can be tested for allowability. If all pass the test, then

$$\mathbf{b}(j) = f_{\mathbf{v}(j)}(\mathbf{b}(j-1)) = \mathbf{b}(j-1) + \mathbf{v}(j),$$

since $\mathbf{b}(j) \in B$ and $\mathbf{v}(j) \in T$, and the conclusion is that the sequence of $\mathbf{v}(j)$'s produces the final balance vector $\mathbf{b}^{(f)}$, which is therefore an acceptable final balance.

The preceding algorithm is to be applied to each of the at most ℓ sequences $\mathbf{v}(1), \ldots, \mathbf{v}(k)$. If a sequence appears for which all the intermediate balances are allowable, then $\mathbf{b}^{(f)}$ is an acceptable final balance vector and the algorithm terminates: if none of the sequences meets this condition, then $\mathbf{b}^{(f)}$ is not a possible final balance. □

The limitations of the algorithm of 8.3.5 will be apparent. It may require enumeration of as many as $|T|^{|B|-1}$ sequences of allowable transaction vectors, a number that is exponential in $|B|$. Of course we could ignore the intermediate balance vectors and just verify that the final balance vector is allowable. In that case the efficient integer programming algorithm employed in 8.3.3 can be applied. One might argue that this weaker verification procedure is sufficient since the intermediate balances are, after all, transient. However, what could not be detected in this way is a transaction which is illegal because of some violation of the intermediate balance restrictions, but which leads to an acceptable final balance.

8.4. More Algorithms

The final section of the chapter is concerned with the construction of algorithms which can decide if a given finitely generated accounting system is one of certain special types discussed in Chapter 7.

(8.4.1). *There are algorithms which, when a finitely generated unbounded accounting system is given, can decide if the system is:*

1. *an inverse system;*

2. *elementary.*

Proof
Let $\mathcal{A}$ be the given system, which is assumed to have the form $(A|\,T)$ with $T = \{\mathbf{v}(1),\ldots,\mathbf{v}(m)\}$, a finite set of allowable vectors; thus $\mathrm{Mon}(\mathcal{A}) = \mathrm{Mon}\langle T\rangle = M$, say.

1. By definition $\mathcal{A}$ is an inverse system if and only if $-\mathbf{v} \in M$ whenever $\mathbf{v} \in M$. Suppose that $\mathbf{v} \in M$ and write $\mathbf{v} = \ell_1 \mathbf{v}(1) + \cdots + \ell_m \mathbf{v}(m)$ where the ℓ_i are non-negative integers. Then

$$-\mathbf{v} = \ell_1(-\mathbf{v}(1)) + \cdots + \ell_m(-\mathbf{v}(m)),$$

from which it follows that $\mathcal{A}$ is inverse if and only if $-\mathbf{v}(i) \in M$ for $i = 1, 2, \ldots, m$. By the membership problem for M – see 8.3.1 and

8.3.3 – we can decide whether $-\mathbf{v}(1), \ldots, -\mathbf{v}(m)$ all belong to M. Therefore we can decide if $\mathcal{A}$ is inverse.

2. In deciding whether $\mathcal{A}$ is elementary, the first step is to identify the set T_0 of all elementary vectors in M. This can be done by testing each of the $n(n-1)$ elementary vectors $\mathbf{e}(i,j)$ for membership in M, where n is the number of accounts. Certainly $\mathrm{Mon}\langle T_0 \rangle \subseteq M$ and M will be elementary if and only if $M \subseteq \mathrm{Mon}\langle T_0 \rangle$. We can test each $\mathbf{v}(i)$ for membership in $\mathrm{Mon}\langle T_0 \rangle$ by 8.3.3. Thus we can decide if $M = \mathrm{Mon}\langle T_0 \rangle$. $\square$

It is also possible to design an algorithm to test an accounting system for the property of being hereditary. However, since this property is best recognized from the feasible digraph, we first need a way to get hold of the digraph. This is accomplished in the next two results.

(8.4.2). *Let $\mathcal{A}$ be a finitely generated, unbounded accounting system. Then there is an algorithm which, when given distinct accounts a_i, a_j, can decide whether $\langle a_i, a_j \rangle$ is an edge of the feasible digraph of $\mathcal{A}$.*

Proof
Let $\mathbf{v}(1), \ldots, \mathbf{v}(k)$ generate $M = \mathrm{Mon}(\mathcal{A})$. Recall that $\langle a_i, a_j \rangle$ is an edge of the feasible digraph if and only there exists a $\mathbf{v} \in M$ such that $v_i < 0$ and $v_j > 0$. Write $\mathbf{v} = \ell_1 \mathbf{v}(1) + \cdots + \ell_k \mathbf{v}(k)$ where the ℓ_r are non-negative integers. Then $\langle a_i, a_j \rangle$ is an edge of the digraph if and only if it is possible to solve the two inequalities

$$\ell_1(v(1))_i + \cdots + \ell_k(v(k))_i < 0 \quad \text{and} \quad \ell_1(v(1))_j + \cdots + \ell_k(v(k))_j > 0$$

for non-negative integers ℓ_r. This an integer program containing inequalities; the standard integer programming algorithms still apply, so it can be determined if there is a non-negative integral solution for the ℓ_r. $\square$

Corollary. *There is an algorithm which, when a finitely generated, unbounded accounting system is given, is able to construct the feasible digraph of the system.*

Proof
Suppose that the system has n accounts. To construct the feasible digraph, test each of the $n(n-1)$ potential edges between vertices for membership in the digraph, using 8.4.2. $\square$

We remark that 8.4.2 and its corollary remain true for finitely specifiable systems, as can be seen by treating allowable transaction types
in the same way as in the proof of 8.3.3.

(8.4.3). *There is an algorithm which can decide whether a given
finitely generated, unbounded accounting system is hereditary.*

Proof
The first step is to decide if $\mathcal{A}$ is elementary, using 8.4.1. Since
hereditary systems are elementary, we may suppose that this is the
case. Next apply the corollary to 8.4.2 to construct the feasible
digraph D of $\mathcal{A}$. For each edge $\langle a_j, a_i \rangle$ of D, we can check to see
if $\mathbf{e}(i,j)$ is feasible, using 8.3.1 and 8.3.3. According to 7.3.6, the
system $\mathcal{A}$ is hereditary if and only if this is true for every edge of
D. It follows that the algorithm can tell if $\mathcal{A}$ is hereditary.   □

Example (8.4.1).

Let $\mathcal{A}$ be the unbounded system with accounts a_1, a_2, a_3 and allowable transactions

$$\mathbf{e}(2,1), \ \mathbf{e}(1,3), \ \mathbf{e}(3,1), \ \mathbf{e}(3,4).$$

Let us test this system to see if it is hereditary. It is obviously an
elementary system. Now construct the feasible digraph D of $\mathcal{A}$,

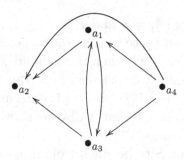

For each edge $\langle a_j, a_i \rangle$ in D, we must verify that $\mathbf{e}(i,j)$ is feasible in $\mathcal{A}$.
This is obviously true except for the edges $\langle a_4, a_1 \rangle, \langle a_4, a_2 \rangle, \langle a_3, a_2 \rangle$;
the equations $\mathbf{e}(1,3) + \mathbf{e}(3,4) = \mathbf{e}(1,4)$, $\mathbf{e}(2,1) + \mathbf{e}(1,3) + \mathbf{e}(3,4) =$
$\mathbf{e}(2,4)$ and $\mathbf{e}(2,1) + \mathbf{e}(1,3) = \mathbf{e}(2,3)$ tell us that the condition holds
for these edges. Therefore $\mathcal{A}$ is hereditary.

Example (8.4.2).

Let $\mathcal{A}$ be the unbounded system with three accounts and three allowable transactions

$$\begin{bmatrix} 100 \\ -200 \\ 100 \end{bmatrix}, \quad \begin{bmatrix} -300 \\ -100 \\ 400 \end{bmatrix}, \quad \begin{bmatrix} -30 \\ -30 \\ 60 \end{bmatrix}.$$

For this system one can see directly that there are no feasible elementary transactions. The reason is that all entries of the allowable vectors are divisible by 10 and hence 10 divides each entry of a feasible transaction, which excludes all the elementary transactions. Consequently $\mathcal{A}$ is not elementary and so it is not hereditary.

It is more of a challenge to construct an algorithm to decide if a finitely generated, unbounded accounting system is simple, i.e., if its monoid can be generated by transaction vectors of the form $m\mathbf{e}(i, j)$ where $m \in \mathbb{Z}$. The final result in the chapter confirms the existence of such an algorithm.

(8.4.4). *There is an algorithm which, when a finitely generated unbounded accounting system $\mathcal{A}$ is given, decides if the system is simple and, if this is the case, finds a finite set of simple transactions that generate the monoid of $\mathcal{A}$.*

Proof
Let $\mathcal{A} = (A|\, T)$ where $|A| = n$ and $T = \{\mathbf{v}_1, \ldots, \mathbf{v}_k\}$ is the finite set of allowable transaction vectors; thus the $\mathbf{v}_i$ generate $M = \mathrm{Mon}(\mathcal{A})$. What must be decided is whether the monoid M can be generated by finitely many simple transactions, i.e., transactions of the form $a_{ij}\mathbf{e}(i, j)$ where $i \neq j$, $1 \leq i, j \leq n$ and the a_{ij} are natural numbers; furthermore, if this is true, it must also be shown how to construct such simple transactions.

Fix $i \neq j$ and define S to be the set of all natural numbers a such that $a\mathbf{e}(i, j) \in M$; then

$$M \cap \mathrm{Mon}\langle \mathbf{e}(i, j)\rangle = \{a\mathbf{e}(i, j)|\, a \in S\},$$

since $\mathrm{Mon}\langle \mathbf{e}(i, j)\rangle$ consists of all vectors of the form $a\mathbf{e}(i, j)$. It is obvious that S is a submonoid of the monoid of natural numbers $\mathbb{N}$, so by 7.2.5 it is finitely generated. Observe that membership in M is decidable by 8.3.1 and 8.3.3. The main step in the proof consists in showing how to construct a finite set of monoid generators for S; this is accomplished in three stages.

(i) *There is an algorithm which, when a positive integer d is given, decides whether $S \subseteq \mathrm{Mon}\langle d \rangle$, and if this is not true, produces an element in $S \backslash \mathrm{Mon}\langle d \rangle$.*

We can assume that $d > 1$. Suppose that $S \not\subseteq \mathrm{Mon}\langle d \rangle$; then there exists an $a \in S$ which is not divisible by d and therefore has the form $a = dq + r + 1$ where d, r are natural numbers and $r < d - 1$. Hence there is an expression

$$(dq + r + 1)\mathbf{e}(i, j) = \ell_1 \mathbf{v}_1 + \cdots + \ell_k \mathbf{v}_k$$

where the ℓ_i are natural numbers. For each r, this vector equation is equivalent to a linear system over $\mathbb{Z}$ in the unknowns q, ℓ_i. Conversely, if there is a solution in non-negative integers q, ℓ_i to the above system for some r where $0 \leq r < d - 1$, then the integer $a = dq + r + 1$ belongs to S, but not to $\mathrm{Mon}\langle d \rangle$ since d does not divide a. It follows that $S \not\subseteq \mathrm{Mon}\langle d \rangle$ if and only if there is a non-negative integer solution of one of the above linear systems for some r where $0 \leq r < d - 1$. By the integer programming algorithm we can decide if such a solution exists and if so, find one.

(ii) *There is an algorithm which finds a finite set of generators for the submonoid S.*

The first step is to decide if $S = \{0\}$. Now $S \neq \{0\}$ if and only if there is an integer $p \geq 0$ such that $(p + 1)\mathbf{e}(i, j) \in M$, i.e.,

$$(p + 1)\mathbf{e}(i, j) = m_1 \mathbf{v}_1 + \cdots + m_k \mathbf{v}_k$$

where the m_i are natural numbers. This is equivalent to a linear system over $\mathbb{Z}$ to be solved for the non-negative integers $p, m_1, \ldots, m_k$. Now we can decide if a solution exists and if so, find one. Thus we can decide whether $S = \{0\}$: of course, should this be the case, nothing more need be done. Therefore we may suppose that $S \neq \{0\}$ and that an element $a_1 \neq 0$ in S has been found; thus $\mathrm{Mon}\langle a_1 \rangle \subseteq S$.

Next we decide whether $S \subseteq \mathrm{Mon}\langle a_1 \rangle$, using (i). If this is true, then $S = \mathrm{Mon}\langle a_1 \rangle$ and we are done. Thus it can be assumed that this containment does not hold and that we have found an element $a_2 \in S \setminus \mathrm{Mon}\langle a_1 \rangle$; then $\mathrm{Mon}\langle a_1, a_2 \rangle \subseteq S$. Denote by d_2 the greatest common divisor of a_1, a_2. Since a_1 does not divide a_2, we have $d_2 < d_1 = a_1$. Next decide if $S \subseteq \mathrm{Mon}\langle d_2 \rangle$. Suppose this is true;

since a_1/d_2 and a_2/d_2 are relatively prime, 7.2.4 may be applied to show that $\mathrm{Mon}\langle d_2\rangle \backslash \mathrm{Mon}\langle a_1, a_2\rangle$ is a finite set and to find an upper bound for its elements. Each non-negative integer not exceeding this bound can be tested for membership in S and any elements of S found in this way may be adjoined to a_1 and a_2 to produce a finite set of generators for S.

Suppose, on the other hand, that $S \not\subseteq \mathrm{Mon}\langle d_2\rangle$; then we can find an element a_3 in S which is not divisible by d_2. Writing d_3 for the greatest common divisor of a_1, a_2, a_3, we have $d_3 < d_2 < d_1 = a_1$ and also $\mathrm{Mon}\langle a_1, a_2, a_3\rangle \subseteq S$. The next step is to decide whether $S \subseteq \mathrm{Mon}\langle d_3\rangle$, and so on.

Since this procedure cannot continue for more than a_1 steps, we will eventually find an integer r for which $d_r = 1$, i.e., the integers $a_1, a_2, \ldots, a_r$ are relatively prime, and of course $\mathrm{Mon}\langle a_1, a_2, \ldots, a_r\rangle \subseteq S$. Therefore, by 7.2.4 again, the set $\mathbb{N}\backslash\mathrm{Mon}\langle a_1, a_2, \ldots, a_r\rangle$ is finite. By testing each of its finitely many elements for membership in S and adjoining any that are found to $\{a_1, a_2, \ldots, a_r\}$, we obtain a finite set of generators for S.

(iii) *Conclusion.*

For each pair of distinct integers i, j in the range $1, 2, \ldots, n$, put $S_{ij} = \{a \in \mathbb{N}|\ a\mathbf{e}(i, j) \in M\}$, so that

$$M \cap \mathrm{Mon}\langle \mathbf{e}(i, j)\rangle = \{a\mathbf{e}(i, j)|\ a \in S_{ij}\}.$$

(Note that S_{ij} might be zero). By (ii) we can find a finite set of generators for each submonoid S_{ij}, say $a_p^{(ij)}, p = 1, \ldots, \ell_{ij}$. Define M_0 to be

$$\mathrm{Mon}\langle a_p^{(ij)}\mathbf{e}(i, j) \mid i, j = 1, 2, \ldots, n,\ i \neq j,\ p = 1, 2, \ldots, \ell_{ij}\rangle,$$

the submonoid generated by all the simple transaction vectors in M; thus $M_0 \subseteq M$. Then $\mathcal{A}$ is simple if and only if $M \subseteq M_0$, i.e., if $\mathbf{v}_i \in M_0$ for $i = 1, 2, \ldots, k$. By 8.3.3 this is decidable, so the algorithm succeeds, i.e., it is able to decide if $\mathcal{A}$ is simple, and in the event that this is true, it constructs a finite set of simple transactions that generate $\mathrm{Mon}(\mathcal{A})$. $\square$

It is worthwhile restating what has just been established. It is possible to determine if a given finitely generated, unbounded accounting system $\mathcal{A}$ is simple and thus if there is a finite set of simple transactions which generate its monoid. Should $\mathcal{A}$ be simple, the algorithm allows us to construct an accounting system equivalent to $\mathcal{A}$ whose allowable transactions are simple. This means that it is possible in principle to "redesign" the accounting system $\mathcal{A}$ in such a way that all the allowable transaction vectors are simple.

Chapter Nine

The Extended Model

9.1. Introduction to the 10-Tuple Model

In any practical accounting system one would expect to find built-in procedures designed to preserve the integrity of the system. The algebraic model described in previous chapters is already equipped with some such procedures: for example, transactions can be screened for allowability before being applied and the resulting balances can be scrutinized. An additional feature of finitely specifiable systems is the capacity to verify final balance vectors and detect improper usage of the system, as was described in Chapter 8.

In this chapter it is shown how to attach two further security mechanisms to the basic model. The first of these is designed to block unauthorized use of the system by verifying that all necessary authorizations have been obtained before a transaction is applied. Typically such authorizations must be obtained from several units of the company, possibly in a specified order. It turns out that such an authorization scheme can be conveniently encoded in two integer matrices called *control matrices*. These matrices are to be attached to the basic model.

Another security mechanism that may be desirable is one that monitors the frequency of application of a particular allowable transaction during an accounting period. The object is to ensure that regular transactions, such as payments on a mortgage, interest on a debt, payment of taxes, etc., are not applied more frequently than is called for. This can be accomplished by specifying a *frequency function* which encodes the number of times that each specific allowable transaction can be applied to the system during an accounting period.

When these mechanisms are adjoined to the basic model, we obtain an extended model of an accounting system which is encoded as a 10-tuple, consisting of sets, functions, vectors and matrices. This *10-tuple model* has many of the capabilities of a real life accounting system. Moreover its operations can be performed by automata which are enhanced versions of the devices introduced in Chapter 6. In addition the model has the advantage that the procedures embedded in the system are, in principle at least, implementable in a standard programming language. Chapter 10 contains a detailed example of the accounting system of a small company, with its 10-tuple extended model.

9.2. Authorization and Control Matrices

Consider the problem of restricting use of an accounting system to authorized units or individuals by requiring that a transaction be authorized according to some prescribed protocol. Suppose that we are dealing with the accounting system of an organization which has a number of divisions, each of which may have subdivisions, departments and so on. The organization can be pictured as a hierarchy in which the smallest independent subdivisions, or *units*, appear at the lowest level. Let the set of units of the firm be ordered in some manner, say as

$$U = \{u_1, u_2, \ldots, u_m\}.$$

Each account in the system will likely be under the control of one or more units, and before a transaction affecting an account can be executed, authorizations must be obtained from the relevant controlling units. In addition, such authorizations may need to be obtained in a particular order. As a further complication, we allow the possibility that different sequences of authorizations for an individual account may be needed according to whether the transaction credits or debits the account. It is this set of protocols governing use of the system that we seek to encode as part of the specification of the system. This can be done conveniently by two integer valued matrices.

Suppose that the accounting system has n accounts $\{a_1, \ldots, a_n\}$. The control mechanism is described by two $n \times m$ matrices with non-negative integer entries

$$C^+ \quad \text{and} \quad C^-.$$

Here the rows of the matrices are labeled by the accounts and the columns by the units of the firm. These matrices are required to have the following property.

$\mathcal{C}$:   *if a row of C^+ or C^- contains an entry $k > 1$, then it also has $k - 1$ as an entry.*

A matrix with this property will be called a *control matrix*. Notice that, as a consequence of the definition, either the ith row of a control matrix consists entirely of zeros or else it contains positive integers $1, 2, \ldots, k_i$, each of which may occur more than once, and possibly some zeros. A little experimentation will show that there are many matrices of this type: an exact count of them will be given later.

The mode of operation of the control matrices

It is now time to explain how the control matrices prevent unauthorized transactions from being applied to the accounting system. Let these matrices be

$$C^+ = [c_{ij}^+]_{n,m} \quad \text{and} \quad C^- = [c_{ij}^-]_{n,m} \; ,$$

which are assumed to have the property $\mathcal{C}$ stated above. Suppose that a transaction $\mathbf{v}$ is to be applied to the system. If $v_i > 0$, i.e., the transaction debits the ith account, then the ith row of matrix C^+ specifies which units must provide authorization for the transaction. Let the non-zero entries in row i of C^+ be

$$1 = c_{ij_1}^+ \leq c_{ij_2}^+ \leq \cdots \leq c_{ij_{r(i)}}^+,$$

where the integers j_ℓ are distinct; then the units which must provide authorizations for the transaction are

$$u_{j_1}, u_{j_2}, \ldots, u_{j_{r(i)}}$$

in that order. What this means is that the numerical order of the non-zero entries of row i determines the sequence of unit authorizations required for the ith account. Notice that, as a consequence of the property $\mathcal{C}$, the sequence $c_{ij_1}^+, c_{ij_2}^+, \ldots, c_{ij_{r(i)}}^+$ consists of the integers $1, 2, \ldots, c_{ij_{r(i)}}^+$ in order *with repetitions allowed.*

What the control matrix C^+ requires is that, if $c_{ij_\ell}^+ < c_{ij_{\ell+1}}^+$, then authorization must be obtained from unit u_{j_ℓ} before unit $u_{j_{\ell+1}}$. If, however, $c_{ij_\ell}^+ = c_{ij_{\ell+1}}^+$, then both u_{j_ℓ} and $u_{j_{\ell+1}}$ must provide authorization, but the order in which this is done is immaterial.

The control matrix C^- operates in a similar fashion. If $v_i < 0$, so that the transaction $\mathbf{v}$ credits the ith account, and if the non-zero entries in row i of C^- are

$$1 = c_{ik_1}^- \le c_{ik_2}^- \le \cdots \le c_{ik_{s(i)}}^-,$$

with distinct integers k_ℓ, then the sequence of units that are required to authorize the transaction is

$$u_{k_1}, u_{k_2}, \ldots, u_{k_{s(i)}}$$

in that order.

Finally, if $v_i = 0$, so that the transaction $\mathbf{v}$ does not affect the ith account, then no authorization is needed and reference to C^+ or C^- is unnecessary. The procedure just described must be applied to all accounts affected by the transaction in question.

In some cases no authorization may be necessary for a transaction to be applied to a particular account, in which event the corresponding rows of C^+ and C^- have zero entries. If some authorizations for an account are required, but in no particular order, then all entries of the corresponding row are 0 or 1. From these examples it is seen that control matrices are a flexible tool for representing complex authorization schemes.

Example (9.2.1).

Consider an organization with three divisions α, β, γ, with γ the accounts department. Suppose α has two subdivisions α_1, α_2 and β has three subdivisions $\beta_1, \beta_2, \beta_3$. Thus in all there are six units, which will be ordered as

$$\alpha_1, \alpha_2, \beta_1, \beta_2, \beta_3, \gamma.$$

For simplicity assume that the organization has just four accounts

$$a_1, a_2, a_3, a_4.$$

The authorizations needed for a transaction to be applied to the system are encoded in the 4×6 control matrices

$$C^+ = \begin{bmatrix} 1 & 1 & 0 & 0 & 0 & 0 \\ 0 & 0 & 0 & 0 & 0 & 0 \\ 1 & 2 & 0 & 3 & 0 & 0 \\ 0 & 0 & 0 & 1 & 0 & 1 \end{bmatrix} \quad \text{and} \quad C^- = \begin{bmatrix} 1 & 0 & 0 & 2 & 0 & 3 \\ 0 & 0 & 0 & 0 & 0 & 1 \\ 0 & 0 & 0 & 0 & 2 & 1 \\ 2 & 0 & 0 & 0 & 0 & 1 \end{bmatrix}.$$

For example, a transaction that debits account a_2 requires no authorization since row 2 of C^+ consists entirely of zeros, while one that credits a_2 must be approved only by the accounts department γ. A transaction that debits a_1 must be authorized by α_1 and α_2, in any order. A transaction that credits a_1 has to be approved by α_1, β_2 and the accounts department γ, in that precise order.

The authorization process

Let us now examine in detail how the authorization process functions for an accounting system $\mathcal{A} = (A \mid T_0, T_1 \mid B)$ with n accounts, allowable transaction types T_0, specific allowable transaction vectors T_1 and allowable balance set B. Suppose that a transaction $\mathbf{v}$ is to be applied. The first step would be to determine if $\mathbf{v}$ is allowable, i.e., if $\mathbf{v} \in T_1$ or type($\mathbf{v}$) $\in T_0$, and then if $\mathbf{v}$ produces an allowable balance, i.e., one in B. Let us assume that $\mathbf{v}$ has already passed these tests.

The next step is to verify that the transaction has received all the necessary authorizations. Let C^+ and C^- be the control matrices which govern authorization of transactions in $\mathcal{A}$. The transaction vector $\mathbf{v}$ will have been approved by certain units of the organization, in a sequence which can be encoded in two further control matrices; thus we can think of $\mathbf{v}$ as being "tagged" by two $n \times m$ control matrices M^+ and M^-, where n and m are the respective numbers of accounts and units in the organization. These matrices record the authorizations which have already been received for the transaction. The system therefore receives an input

$$(\mathbf{v}, M^+, M^-).$$

The role of the matrices M^+ and M^- must now be explained. Suppose first that $v_i > 0$ and that

$$1 = m^+_{ij_1} \leq m^+_{ij_2} \leq \cdots \leq m^+_{ij_{t(i)}}$$

are the non-zero entries in row i of M^+. This means that authorizations to debit the balance of account a_i have already been provided by units $u_{j_1}, u_{j_2}, \ldots, u_{j_{t(i)}}$, in that order. In addition it is understood that if $v_i \leq 0$, then row i of M^+ consists of zeros: this is because the matrix M^+ is only relevant to checking authorizations of debits. Next suppose that $v_i < 0$; then there is a similar interpretation of the role of the entries in row i of M^- regarding authorizations obtained for crediting accounts. Again rows of M^- for which $v_i \geq 0$ are all zero.

Before the transaction $\mathbf{v}$ can be approved, the matrices M^+, M^- must be compared with C^+, C^-. If $v_i > 0$, then m_{ij}^+ must equal c_{ij}^+, provided that $c_{ij}^+ \neq 0$. If $v_i < 0$, the condition is that $m_{ij}^- = c_{ij}^-$, provided that $c_{ij}^- \neq 0$. If $v_i = 0$, there is no condition since the transaction does not affect account a_i.

There is a convenient symbolic way of expressing the relationship that must hold between the matrices M^+, C^+ and M^-, C^-. For a given n-column vector $\mathbf{v}$, we define a relation $>_{\mathbf{v}}$ between nonnegative, integer valued $n \times m$ matrices X, Y as follows:

$$X >_{\mathbf{v}} Y$$

is to mean that if $v_i > 0$, then $x_{ij} = y_{ij}$ whenever $y_{ij} \neq 0$. The verification that the transaction $\mathbf{v}$ has received all the necessary approvals can then be written in the matrix form

$$M^+ >_{\mathbf{v}} C^+ \text{ and } M^- >_{-\mathbf{v}} C^-.$$

We note that the ith row of M^+ or M^- will be zero if $v_i \leq 0$ or $v_i \geq 0$ respectively, because no authorizations are required in these cases. Since in practice most transactions affect few accounts, the matrices M^+ and M^- will consist largely of zeros. Thus for economy of display and storage it is desirable not to list these zero rows; therefore in specific examples we will delete any row of M^+ or M^- for which $\mathbf{v}$ has an entry which is not positive or not negative respectively, and work with the resulting *reduced* matrices

$$M^{+*} \text{ and } M^{-*}.$$

Notice that the matrices M^+ and M^- can be reconstructed from knowledge of M^{+*}, M^{-*} and the vector $\mathbf{v}$: for example, if $v_i \leq 0$, we insert a row of zeros as the ith row of M^{+*}, with a similar procedure for the matrix M^{-*} if $v_i \geq 0$. The condition on matrices can therefore be stated unambiguously, if with some abuse of notation, in the form

$$M^{+*} >_{\mathbf{v}} C^+ \text{ and } M^{-*} >_{-\mathbf{v}} C^-.$$

Described in words, the verification process to authorize a transaction $\mathbf{v}$ is as follows. For $i = 1, 2, \ldots, n$, if $v_i > 0$, row i of M^+

is compared with row i of C^+; for each positive entry in C^+ there should be the same entry in M^+, but M^+ might have further non-zero entries in the row if additional authorizations beyond those that are strictly necessary have been obtained. If $v_i < 0$, the positive entries of row i of C^- are compared with those of M^- in the same manner. If $v_i = 0$, no authorizations for account i are necessary and no comparisons need be made.

These matrix comparisons are to be performed for each row. If they are all performed satisfactorily, the approval process for **v** is complete and the transaction is fully authorized. If, on the other hand, the comparison fails for any account, the transaction will be rejected as not being properly authorized.

The number of control matrices

As one would expect, there are many control matrices of given size, although probably only a few of them would be used in practice. We pause to show that an exact count of control matrices is possible.

(9.2.1). *The number of $n \times m$ control matrices is*

$$\left(1 + \sum_{k=1}^{m} \sum_{p=1}^{k} p!\, S(k,p) \binom{m}{k}\right)^n$$

where the $S(k,p) = \frac{1}{p!} \sum_{i=0}^{p-1} (-1)^i \binom{p}{i} (p-i)^k$ are the Stirling numbers of the second kind.

Proof
It is enough to establish the formula in the case when $n = 1$, i.e., there is a single row, since the general result will then follow by raising the result to the nth power.

One possible row is the row of m zeros. We need to count the non-zero rows. Consider a row with exactly k non-zero entries where $1 \leq k \leq m$. First choose the positions in the row which are to receive non-zero entries in $\binom{m}{k}$ ways. Then count the number of ways to fill the k chosen positions with positive integers, subject to the condition C on the rows of a control matrix. Suppose that p is the largest positive integer which actually appears in the row; then $1 \leq p \leq k$ since each positive integer less than p must also occur in the row. The number of ways to fill the k positions is equal to the number of ways to place k distinct objects in p distinct boxes in order with at least one element in each box: the last requirement is needed to ensure that each of the integers $1, 2, \ldots, p$ appears at least once in the row. This is yet another distribution problem. It is well known from combinatorics this number is equal to $p!S(k,p)$ – see [2]. Therefore the number of possible rows is

$$1 + \sum_{k=1}^{m} \sum_{p=1}^{k} p!S(k,p) \binom{m}{k}.$$

$\square$

For example, if the organization has just three units, the above formula gives the number of possible rows as 26, so that, if there are n accounts, the number of control matrices is $(26)^n$.

9.3. Frequency Control

Another security device which can be incorporated in the basic model of an accounting system is a mechanism to control the frequency with which a particular allowable transaction is applied during an accounting period. Without such a device there might be nothing to prevent an allowable transaction which has been fully authorized from being applied with greater frequency than permitted by company rules: for example, this would apply to regularly scheduled transactions. We aim to show that the frequency of application of a transaction can be monitored by means of a so-called frequency function.

Consider an accounting system

$$\mathcal{A} = (A|\ T_0,\ T_1),$$

where T_0 and T_1 are the sets of allowable transaction types and allowable specific transactions respectively. The objective is to monitor the number of applications of a specific transaction in T_1 over an accounting period. For this purpose a function

$$\varphi : T_1 \to \mathbb{N} \cup \{\infty\}$$

is introduced, the idea being that a given transaction $\mathbf{v} \in T_1$ may not be applied more than $\varphi(\mathbf{v})$ times. If $\varphi(\mathbf{v}) = \infty$, then it is to be understood that there is no limitation on the number of times that $\mathbf{v}$ can be applied during the period. If $\varphi(\mathbf{v}) = 0$, then $\mathbf{v}$ cannot be applied during this time: it is useful to allow this possibility since, for example, it might be necessary to suspend a regular payment during a certain time period, which might be preferable to eliminating it altogether from the system as allowable transaction. We refer to such a function φ as a *frequency function*. If the set T_1 is ordered in some fixed manner, then φ can be conveniently identified with column vector over $\mathbb{N} \cup \{\infty\}$ with $|T_1|$ rows, namely the values of the function φ.

Let us see how the frequency function operates. Let $\mathbf{v} \in T_1$ be an allowable transaction vector and assume it has been fully authorized and that it leads to an acceptable balance vector. At any instant there is a *frequency counter*, by which we mean a function

$$\kappa : T_1 \to \mathbb{N}$$

such that $\kappa(\mathbf{v})$ is the number of times the transaction $\mathbf{v}$ has already been applied during the accounting period. If $\kappa(\mathbf{v}) < \varphi(\mathbf{v})$, then the transaction may be applied and the value of the counter κ at $\mathbf{v}$ is reset to $\kappa(\mathbf{v}) + 1$. However, if $\kappa(\mathbf{v}) = \varphi(\mathbf{v})$, the transaction is rejected, since it has already been applied the maximum permitted number of times. Notice that with these rules it is impossible to have $\kappa(\mathbf{v}) > \varphi(\mathbf{v})$. Thus the frequency counter is adjusted by one each time that a transaction is successfully applied; when the system is regarded as an automaton, the function κ is part of the state of the machine. As in the case of the function φ, we think of κ as a $|T_1|$-column vector, but over $\mathbb{N}$.

9.4. The 10-Tuple Model and Automata

In this section our aim is to adjoin formally the security mechanisms described in 9.2 and 9.3 to the basic model of an accounting system. The resulting extended model has the ability to simulate many features of a realistic accounting system and it can be represented by enhanced versions of the automata described in Chapter 6.

We begin with the basic model of a bounded accounting system over an arbitrary ordered domain R

$$\mathcal{A} = (A \mid T_0, T_1 \mid \lambda, \Lambda),$$

where $A = \{a_1, \ldots, a_n\}$ is the set of accounts, T_0 and T_1 are the respective sets of allowable transaction types and specific allowable transactions, and

$$\lambda, \Lambda : A \to R \cup \{-\infty, \infty\}$$

are bounding functions for account values: thus $\lambda(a_i) \leq \Lambda(a_i)$ for $i = 1, 2, \ldots, n$, and the set of allowable balance vectors is

$$B = \{\mathbf{v} \in \mathrm{Bal}_n(R) |\ \lambda(a_i) \leq v_i \leq \Lambda(a_i),\ i = 1, 2, \ldots, n\}.$$

Note that λ and Λ may be regarded as n-column vectors with entries in $R \cup \{-\infty, \infty\}$.

In addition the extended model is to have a built-in capacity to generate reports. Recall from Chapter 5 that a report corresponds to an equivalence relation E on the account set A. The balance vector of the quotient system $\mathcal{A}/E$ gives basic information about the report, namely the list of balances of accounts in each equivalence class. Thus the accounting system $\mathcal{A}$ should be equipped with a set

$$\mathcal{E}$$

of equivalence relations on A which generate all the necessary reports. Now an equivalence relation E on A is specified by an $n \times n$ matrix which decomposes into blocks of 0's and 1's: for the (i, j)th matrix entry equals 1 precisely when $a_i\ E\ a_j$ and is otherwise 0, so that each E-equivalence class corresponds to a square submatrix of 1's. Thus we can regard $\mathcal{E}$ as a set of matrices of this type.

Next assume that the organization to which the accounting system $\mathcal{A}$ belongs consists of m autonomous units, written in the fixed order $u_1, u_2, \ldots, u_m$, and write

$$U = \{u_1, u_2, \ldots, u_m\}.$$

Each account in A is governed by certain units which must authorize transactions affecting the account. The sequences of authorizations required are encoded in two $n \times m$ control matrices

$$C^+ \quad \text{and} \quad C^-,$$

as described in 9.2: recall that these are $n \times m$ matrices with non-negative entries which have the property C: if an integer $\ell > 1$ appears in a row, then so does $\ell - 1$.

Finally, a frequency function

$$\varphi : T_1 \to \mathbb{N} \cup \{\infty\}$$

is introduced to control the frequency of application of each specific allowable transaction: φ is identified with a $|T_1|$-column with entries in $\mathbb{N} \cup \{\infty\}$.

The result of these additions to the basic model is a 10-tuple which will be called *the extended model* of an accounting system,

$$\mathcal{A} = (A \mid T_0, T_1 \mid \lambda, \Lambda \mid \mathcal{E} \mid U, C^+, C^- \mid \varphi).$$

Features of the extended model

We summarize the capabilities of the 10-tuple model displayed above, where now it is assumed that the ordered domain is $\mathbb{Z}$.

1. The system is able to generate reports corresponding to equivalence relations E in the set $\mathcal{E}$ by passing to the quotient system $\mathcal{A}/E$. It may be convenient to include in $\mathcal{E}$ the trivial equivalence of equality E_0.

2. The current value of the frequency counter κ at $\mathbf{v} \in T_1$ is the number of times that the transaction $\mathbf{v}$ has been successfully applied during the current period.

3. If T_1 is finite, so that $\mathcal{A}$ is finitely specified, it is possible to verify that a final balance could really have been obtained by legitimate actions, when intermediate balance restrictions are not considered. For this purpose the algorithm in 8.3 must be attached to the model; recall that it is based on the integer programming algorithm.

4. If λ and Λ are finite valued, it is possible to verify a final balance vector while taking into account restrictions on intermediate account balances, albeit by a less efficient algorithm.

5. If the set T_0 is finite and the system is unbounded, i.e., $\lambda(a_i) = -\infty$ and $\Lambda(a_i) = +\infty$ for $i = 1, 2, \ldots, n$, then it is possible to determine whether the accounting system is simple or elementary. Moreover, if this is the case, it is possible to find sets of simple or elementary transactions which generate the monoid of $\mathcal{A}$: for these results see 8.4.

The extended automata

The extended model of an accounting system, like the basic model, lends itself to interpretation as an automaton. Recall that there are two automata associated with the basic model: we begin with the one which does not involve time. Let $\mathcal{A}$ be the 10-tuple extended accounting system over an ordered domain R,

$$\mathcal{A} = (A|\, T_0,\, T_1|\, \lambda,\, \Lambda|\, \mathcal{E}|\, U,\, C^+,\, C^-|\, \varphi)$$

with the notation described above. Next we will define the extended automaton of $\mathcal{A}$

$$\mathcal{M_A} = (Z, X, Y, \delta, \lambda).$$

This is to have input set

$$X = \mathrm{Bal}_n(R) \times CM_{n,m} \times CM_{n,m}$$

where $CM_{n,m}$ is the set of all $n \times m$ control matrices, while the state set is

$$Z = \mathrm{Bal}_n(R) \times \mathrm{Fun}(T_1, \mathbb{N}).$$

The output of the original automaton is modified so as to incorporate the capability of generating the reports corresponding to the equivalence relations in $\mathcal{E}$. The output set is taken to be the cartesian product

$$Y = \mathrm{Cr}_{E \in \mathcal{E}}\, \mathrm{Bal}_{n(E)}(R),$$

possibly augmented by error messages: here $n(E)$ is the number of E-equivalence classes.

The change of state function δ is given by the rule

$$\delta((\mathbf{b}, \kappa), (\mathbf{v}, M^+, M^-)) = (\mathbf{b} + \mathbf{v}, \kappa')$$

where κ' is defined by

$$\kappa'(\mathbf{v}) = \kappa(\mathbf{v}) + 1 \quad \text{and} \quad \kappa'(\mathbf{w}) = \kappa(\mathbf{w}), \text{ if } \mathbf{w} \neq \mathbf{v}.$$

This is provided that the following are true: $\mathrm{type}(\mathbf{v}) \in T_0$ or $\mathbf{v} \in T_1$, $\lambda(a_i) \leq b_i + v_i \leq \Lambda(a_i)$ for $i = 1, 2, \ldots, n$, $M^+ >_{\mathbf{v}} C^+$ and $M^- >_{-\mathbf{v}} C^-$, and $\kappa(\mathbf{v}) < \varphi(\mathbf{v})$. However, $\delta((\mathbf{b}, \kappa), (\mathbf{v}, M^+, M^-)) = (\mathbf{b}, \kappa)$ if any one of these conditions fails to hold.

The output function λ is computed from the equation

$$\lambda((\mathbf{b}, \kappa), (\mathbf{v}, M^+, M^-)) = ((\overline{\mathbf{b} + \mathbf{v}})_E)_{E \in \mathcal{E}},$$

provided that type$(\mathbf{v}) \in T_0$ or $\mathbf{v} \in T_1$, $\lambda(a_i) \le b_i + v_i \le \Lambda(a_i)$, $M^+ >_{\mathbf{v}} C^+$ and $M^- >_{-\mathbf{v}} C^-$, and $\kappa(\mathbf{v}) < \phi(\mathbf{v})$: otherwise an error message is printed as the output. Recall the notation from Chapter 5 which is used here: $\overline{\mathbf{b}}_E \in \mathrm{Bal}_{n(E)}(R)$ has as its i-component $\sum_{a_j E a_i} b_j$. Thus the output function produces all the relevant reports after each transaction has been applied. Notice that if $E_0 \in \mathcal{E}$, then the E_0-component of the output is just the final balance vector of the system after the transaction has been applied.

To summarize in words the operation of the automaton $\mathcal{M}_{\mathcal{A}}$, suppose that the input $(\mathbf{v}, M^+, M^-)$ is applied. This means that $\mathbf{v} \in \mathrm{Bal}_n(R)$ and that the authorization sequences already received for the transaction $\mathbf{v}$ are displayed in the control matrices M^+, M^-. Assume that the current state of the system is $(\mathbf{b}, \kappa)$; here $\mathbf{b}$ is the current balance vector and the value of the frequency counter $\kappa : T_1 \to \mathbb{N}$ at $\mathbf{v}$ records the number of times that this transaction has been applied up to this time.

The automaton first determines if type$(\mathbf{v}) \in T_0$ or if $\mathbf{v} \in T_1$: it then determines if the new balance vector $\mathbf{b} + \mathbf{v}$ satisfies $\lambda_i \le b_i + v_i \le \Lambda_i$ for $i = 1, 2, \ldots, n$. If this test is passed, the next step is to verify that $M^+ >_{\mathbf{v}} C^+$ and $M^- >_{-\mathbf{v}} C^-$. This is to check that the transaction $\mathbf{v}$ has received all the required authorizations. The final test is whether $\kappa(\mathbf{v}) < \varphi(\mathbf{v})$ in the case where $\mathbf{v} \in T_1$, i.e., the transaction has not already been applied the maximum permitted number of times.

Should all these verifications be performed satisfactorily, the state of the automaton changes to $(\mathbf{b} + \mathbf{v}, \kappa')$ where $\kappa'(\mathbf{v}) = \kappa(\mathbf{v}) + 1$ and $\kappa'(\mathbf{w}) = \kappa(\mathbf{w})$ if $\mathbf{w} \ne \mathbf{v}$: the output is $((\overline{\mathbf{b} + \mathbf{v}})_E)_{E \in \mathcal{E}}$. However, if any of the tests fail, the state of the system does not change and an appropriate error message is generated as the output.

It might be objected that the output of the automaton is more complex than would normally be required since it includes all the reports after each transaction. This can be avoided by projecting the output onto its E_0-component, thereby giving just the final balance vector of the whole system: however, the facility of producing multiple reports might be useful enough to justify the present form.

The extended time enhanced accounting system

Next we describe the time enhanced automaton of an extended accounting system with n accounts over an ordered domain R in the standard 10-tuple form

$$\mathcal{A} = (A \mid T_0, T_1 \mid \lambda, \Lambda \mid \mathcal{E} \mid U, C^+, C^- \mid \varphi),$$

where for simplicity we have assumed that $\mathcal{E}$ contains a single equivalence relation E. Let $A = \{a_1, a_2, \ldots, , a_n\}$ be the account set, with $A_{re} = \{a_{r+1}, \ldots, a_n\}$ the set of revenue and expense accounts.

The corresponding extended time enhanced automaton is

$$\mathcal{T}_\mathcal{A} = (Z, X, Y, \delta, \lambda),$$

where the parameters are defined as follows.

- The set of states Z is a subset of

$$(\overline{\mathbb{Z} \times R})^n \times \mathrm{Fun}(T_1, \mathbb{N}),$$

where T_1 is the set of specific allowable transactions. Thus a state of $\mathcal{T}_\mathcal{A}$ has the form

$$(\overline{z}_1 \overline{z}_2 \ldots \overline{z}_n, \kappa),$$

where

$$\overline{z}_i = z_{i1} z_{i2} \ldots z_{im(i)}, \quad z_{ij} = (t_{ij}, x_{ij}), \quad (t_{ij} \in \mathbb{Z}, \; x_{ij} \in R),$$

and κ is a function from T_1 to $\mathbb{N}$ which counts the number of times that each allowable transaction has been applied successfully. Here it is understood that

$$t_{i1} \le t_{i2} \le \cdots \le t_{im(i)}$$

and

$$x_{ij} \neq 0 \text{ and } \sum_{i=1}^{n} \sum_{j=1}^{m(i)} x_{ij} = 0.$$

- The set of inputs is

$$X = \mathbb{Z} \times \mathrm{Bal}_n(R) \times CM_{n,m} \times CM_{n,m},$$

where $CM_{n,m}$ is the set of all $n \times m$ control matrices: thus a typical input has the form

$$(t, \mathbf{x}, M^+, M^-) :$$

here $\mathbf{x} \in \mathrm{Bal}_n(R)$ has components $x_1, x_2, \ldots, x_n,$, $t \in \mathbb{Z}$ and M^+ and M^- are control matrices encoding the sequences of authorizations obtained for the transaction from the m autonomous units, which are written in a fixed order $u_1, u_2, \ldots, u_m$.

- The next state function

$$\delta : Z \times X \longrightarrow Z$$

  is defined by the rules that follow.

1. δ maps $((\overline{z}_1\overline{z}_2 \ldots \overline{z}_n, \kappa), (t, \mathbf{x}, M^+, M^-))$ to

$$(\overline{z}_1(t, x_1)\overline{z}_2(t, x_2) \ldots \overline{z}_n(t, x_n), \kappa'),$$

   where $\kappa' : T_1 \to \mathbb{N}$ is such that

$$\kappa'(\mathbf{v}) = \begin{cases} \kappa(\mathbf{x}) + 1, & \text{if } \mathbf{v} = \mathbf{x}, \\ \kappa(\mathbf{v}), & \text{otherwise} \end{cases}$$

   provided that
   - (a) $t \geq t_{ij}$ for $i = 1, 2, \ldots, n$, $j = 1, 2, \ldots, m(i)$;
   - (b) $(\overline{z}_1(t, x_1)\overline{z}_2(t, x_2) \ldots \overline{z}_n(t, x_n), \kappa') \in Z$;
   - (c) $M^+ >_{\mathbf{x}} C^+$ and $M^- >_{-\mathbf{x}} C^-$.

2. δ maps $((\overline{z}_1\overline{z}_2 \ldots \overline{z}_n, \kappa), (t, \mathbf{x}, M^+, M^-))$ to $(\overline{z}_1\overline{z}_2 \ldots \overline{z}_n, \kappa)$, if any of the conditions (a),(b),(c) above fails.

   (Recall that $\overline{z}_i(t, x_i)$ is the concatenation of $\overline{z}_i$ and (t, x_i), except that if $x_i = 0$, the pair (t, x_i) is to be omitted from the sequence).

- The output set is

$$Y = \mathbb{Z} \times \mathrm{Bal}_{\overline{n}}(R),$$

where $\overline{n}$ is the number of accounts in the quotient system $\mathcal{A}/E$. The output function

$$\lambda : Z \times X \longrightarrow Y$$

is defined by having it send $((\overline{z}_1\overline{z}_2 \ldots \overline{z}_n, \kappa), (t, \mathbf{x}, M^+, M^-))$ to the ordered pair $(t, \sigma_E^*(\overline{x}))$ where $\sigma_E : \mathcal{A} \to \mathcal{A}/E$ is the canonical epimorphism and $\overline{x}$ is the balance vector with entries

$$x_1 + \sum_{j=1}^{m(1)} x_{1j}, \; x_2 + \sum_{j=1}^{m(2)} x_{2j}, \; \ldots, \; x_n + \sum_{j=1}^{m(n)} x_{nj}.$$

Here $m(i)$ is the number of entries in the T-diagram of the ith account. On the other hand, if $t < t_{ij}$ for some i, j, then the value of the output function is a suitable error message.

Recall that in the definition of the time enhanced automaton in 6.3 the output function combined the balances of the revenue and expense accounts to produce the net income. Thus an equivalence relation of particular interest is one for which all these accounts are combined. Assume therefore that the set of revenue and expense accounts A_{re} constitutes one equivalence class of the equivalence relation E; then the entry of $\sigma_E^*(\overline{x})$ corresponding to the equivalence class A_{re} is

$$\sum_{i=r+1}^{n} \left(x_i + \sum_{j=1}^{m(i)} x_{ij} \right),$$

which is the net income of the system.

9.5. The Audit as an Automaton

In sections 9.2 and 9.3 control mechanisms were introduced which guarantee that all transactions applied to an accounting system have the required authorizations and that transactions are not applied more than the permitted number of times. However, apart from these *a priori* procedures, a company will be subject to *a posteriori* control mechanisms, depending on its legal status. Thus it is usual that after a certain period of time, generally a year, the balance vector of account balances and the financial activities that have occurred in the system during the period must be checked in order to determine if any procedural errors have occurred during the accounting process. This verification process is carried out by an auditor using the so-called *balance-check tests*. In this section it is shown how to design an automaton which performs the mechanical aspects of the auditor's task.

Typically there are six types of error that occur during the operation of an accounting system.

1. The accounts affected by a transaction resulting from some economic activity are not the appropriate ones.

2. The accounts affected by the economic event are appropriate, but they do not coincide with the accounts approved by the auditors.

3. The economic event has not been registered.

4. The transaction resulting from the economic event does not correspond to a balance vector.

5. The balance vector obtained after application of the transaction is not allowable.

6. A report generated by the transaction is not permissible.

For certain types of error it is the duty of the firm of auditors to communicate to the company the flaws detected in the accounting process. Then the relevant accounts, allowable transactions, allowable balance vectors or set of reports may have to be modified to meet the objections of the auditor.

Assume that over a certain period of time a company operates the accounting system over an ordered domain

$$\mathcal{A} = (A|\ T|\ B),$$

where A is the set of n accounts, T the set of allowable transactions and B the set of allowable balance vectors. Recall that T consists of all vectors of certain allowable types T_0, as well as a set of specific allowable vectors T_1.

At the end of the accounting period it is the task of the auditor to perform certain tests in order to verify that all the economic events affecting the company have been accounted for, that all generally accepted accounting principles and criteria have been applied, and that the accounting system is functioning correctly. In order to accomplish this, the auditor selects a certain equivalence relation E^c on the account set A and chooses a sample set of E^c-equivalence classes of accounts, say

$$\overline{a}_{j_1}, \overline{a}_{j_2}, \ldots, \overline{a}_{j_m},$$

where $1 \leq j_1 < j_2 < \cdots < j_m \leq n$. Usually this sample of accounts includes "bank account", "trade debtors", "inventories" and "trade creditors", where these accounts may be aggregates of individual accounts, as explained below.

- "Bank account" may consist of several accounts held at different institutions.

- "Trade debtors" and "trade creditors" appear separately, but are actually lists of specific debtors and creditors.

- "Inventories" are broken down according to the different categories, typically:

  - Goods for resale.
  - Finished goods.
  - Semi-finished goods.
  - Byproducts and waste.
  - Work in process.
  - Raw materials and supplies.
  - Parts and subassemblies.
  - Consumables and spares.
  - Packing materials and containers.

Thus the $\bar{a}_{j_\ell}$ are actually accounts in the quotient system $\mathcal{A}/E^c$, which is the system that the auditor deals with. If the sequence of balance vector entries for account $\bar{a}_k$, $k \in \{j_1, j_2, \ldots, j_m\}$, is

$$\bar{z}_k = z_{k1} z_{k2} \ldots z_{km(k)} \quad \text{where} \quad z_{k\ell} = (t_{k\ell}, x_{k\ell}), \quad (t \in \mathbb{Z}, \ x_{k\ell} \in R),$$

then in the expression for $\bar{z}_k$ the auditor will have combined in each $x_{k\ell}$ the amounts for all the accounts in the E^c-equivalence class $\bar{a}_k$: for example, the cash balances of the various bank accounts are totalled. (Notice that we are not dealing a true report here since not all accounts in $\mathcal{A}/E^c$ have been selected).

Next the auditor makes a selection from the E^c-equivalence classes $\bar{a}_{j_1}, \bar{a}_{j_2}, \ldots, \bar{a}_{j_m}$. This sample is to be chosen by the statistical technique of stratified sampling. Having selected the equivalence classes, the auditor must verify the final balance of each one by obtaining the balances of all the accounts in the selected equivalence class and combining them. Of course, in order to guarantee the independence of the auditing process, these final balances must be requested not from within the company, but from the external sources, for example, from the banks, debtors, creditors and warehouse.

In case a final balance provided to the auditor by an external unit does not coincide with the final balance generated by the company's accounting process, the auditor must request all the documentation related to the corresponding account. Naturally, this documentation must again be supplied by the external sources. Finally, the auditor revises the documents supplied and decides if the error is due to the company or to the provider (in the case of inventories, it is necessary to use the appropriate inventory count sheets). In the first case, the auditor must propose a correction to the final balance of the company.

In practice, the company is obliged to correct a detected error in the final balance only if the mistake is materially significant. On the other hand, if the errors detected exceed a certain number, which is predicted by statistical methods, namely quality control techniques, the auditor may have insufficient confidence in the company's accounts and might require that further tests be applied.

After completion of this procedure, the auditor will propose a new set of specific allowable transactions, denoted by T_1^c, which will be called the *control set* of allowable transactions,

$$\{\mathbf{v}_1^c, \mathbf{v}_2^c, \ldots, \mathbf{v}_r^c\}.$$

Observe that the balance vectors in T_1^c are specific balance vectors and that the number of elements in T_1^c depends on the method of choosing a representative sample of the suspicious transactions, say

$$\mathbf{v}_1, \mathbf{v}_2, \ldots, \mathbf{v}_r.$$

Note that some vectors $\mathbf{v}_i$ will be $\mathbf{0}$ if the corresponding transactions proposed by the auditor are new.

We now define an *audit* of the accounting system $\mathcal{A} = (A|\,T|\,B)$ to be an automaton

$$\mathcal{M}^c$$

whose parameters are as follows.

- The state set is B.

- The input set is $T_1^c = \{\mathbf{v}_1^c - \mathbf{v}_1, \mathbf{v}_2^c - \mathbf{v}_2, \ldots, \mathbf{v}_r^c - \mathbf{v}_r\}$.

- The output set is $\overline{B}$ where $\overline{B}$ is the quotient set B/E^c obtained from the equivalence relation E^c.

- Changes of state are determined by the next state function $\delta^c : B \times T_1^c \to B$ given by the rule

$$\delta^c(\mathbf{b}, \mathbf{v}_i^c - \mathbf{v}_i) = \mathbf{b} + \mathbf{v}_i^c - \mathbf{v}_i.$$

- The output is computed from the output function
  $\lambda^c : \overline{B} \times T_1^c \to \overline{B}$, which is defined by

$$\lambda^c([\mathbf{b}]_{E^c}, \mathbf{v}_i^c - \mathbf{v}_i) = [\mathbf{b} + \mathbf{v}_i^c - \mathbf{v}_i]_{E^c}.$$

In essence what the auditing automaton $\mathcal{M}^c$ achieves is replacement of suspect specific allowable transactions $\mathbf{v}_i$ by corrected transactions $\mathbf{v}_i^c$. Thus the audit of the accounting system can be thought of as an associated automaton devised by the auditor with the aim of verifying the operation of the original system. It is not a part of the accounting system like the control mechanisms described in 9.2 and 9.3, but is an *a posteriori* device imposed by an external source.

Chapter Ten

The Model Illustrated

The purpose of this final chapter is to illustrate the applicability of the extended model by presenting a detailed account of the accounting system of a small company in terms of the 10-tuple model developed in Chapter 9.

10.1. A Real Life Example

Let us consider the case of a company which is engaged in the business of trading finished products. We aim to show in detail how the operation of the company's accounting system can be represented by the extended model in the form of a standard 10-tuple

$$\mathcal{A} = (A \mid T_0, \ T_1 \mid \lambda, \ \Lambda \mid \mathcal{E} \mid U, \ C^+, \ C^- \mid \varphi).$$

It is assumed that the company has four departments:

α : Cash.

β : Customer order department.

γ : Warehouse.

δ : Accounts department

The customer order department has two subdivisions, namely

β_1 : Purchasing.

β_2 : Sales.

Thus in all there are five units in the company and the set of units is

$$U = \{\alpha, \ \beta_1, \ \beta_2, \ \gamma, \ \delta\}.$$

Next we assume that the accounting system of the company has just 12 accounts and the account set is

$$A = \{a_1, a_2, \ldots, a_{12}\},$$

where the accounts are given by the following key:

a_1 : Cash on hand.

a_2 : Bank account.

a_3 : Trade debtors.

a_4 : Machinery, plant and tools.

a_5 : Buildings and other structures.

a_6 : Inventories.

a_7 : Share capital.

a_8 : Trade creditors.

a_9 : Loans received.

a_{10} : Accumulated depreciation of fixed assets.

a_{11} : Retained earnings.

a_{12} : Profit and loss.

According to the rules and practices of the company, there are six allowable transaction types and three specific allowable transactions. In interpreting the balance vectors that follow the reader is reminded of the convention that debits increase account balances while credits decrease them, after allowing for the signs of the account balances.

The set T_0 of allowable transaction types consists of the following:

- Purchase of inventories, to be paid in part through cash or bank account and in part through trade creditors,

$$
\begin{bmatrix}
- \\
0 \\
0 \\
0 \\
0 \\
+ \\
0 \\
- \\
0 \\
0 \\
0 \\
0
\end{bmatrix}
,
\begin{bmatrix}
0 \\
- \\
0 \\
0 \\
0 \\
+ \\
0 \\
- \\
0 \\
0 \\
0 \\
0
\end{bmatrix}
.
$$

- Sale of inventories, to be paid in part through cash or bank account and in part through trade debtors,

$$
\begin{bmatrix}
+ \\
0 \\
+ \\
0 \\
0 \\
- \\
0 \\
0 \\
0 \\
0 \\
0
\end{bmatrix}
,
\begin{bmatrix}
0 \\
+ \\
+ \\
0 \\
0 \\
- \\
0 \\
0 \\
0 \\
0 \\
0
\end{bmatrix}
.
$$

- Receipt of a bank loan to purchase machinery or inventories,

$$
\begin{bmatrix} 0 \\ 0 \\ 0 \\ + \\ 0 \\ 0 \\ 0 \\ 0 \\ - \\ 0 \\ 0 \\ 0 \end{bmatrix} , \quad \begin{bmatrix} 0 \\ 0 \\ 0 \\ 0 \\ 0 \\ + \\ 0 \\ 0 \\ - \\ 0 \\ 0 \\ 0 \end{bmatrix} .
$$

Next the set T_1 of specific allowable transactions consists of the following:

- w_1: depreciation of the machinery, plant and tools at 5% per year. Assume that the initial balance of account a_4 is \$100,000, so the depreciation is \$5,000 per year:

$$
\begin{bmatrix} 0 \\ 0 \\ 0 \\ 0 \\ 0 \\ 0 \\ 0 \\ 0 \\ 0 \\ -5,000 \\ 0 \\ 5,000 \end{bmatrix} .
$$

- w_2: quarterly loan amortization paid through bank account; the loan amount is assumed to be \$6,000 and rate of interest 1% per trimester; the amount of principal repaid is \$150, so the quarterly payment is $210:

$$\begin{bmatrix} 0 \\ -210 \\ 0 \\ 0 \\ 0 \\ 0 \\ 0 \\ 0 \\ 150 \\ 0 \\ 0 \\ 60 \end{bmatrix}.$$

- w_3: the company has two employees. The remuneration of each one is $1,500 a month payable through bank account and is counted against profit/loss. Moreover, they earn two extra month's salaries paid twice a year:

$$\begin{bmatrix} 0 \\ -3,000 \\ 0 \\ 0 \\ 0 \\ 0 \\ 0 \\ 0 \\ 0 \\ 0 \\ 0 \\ 3,000 \end{bmatrix}.$$

Next the allowable account balances for the company are specified by bounding functions λ, Λ. It is convenient to identify these functions with two 12-column vectors with entries in $\mathbb{Z} \cup \{-\infty, \infty\}$, where the ith component of the vectors are $\lambda(a_i)$ and $\Lambda(a_i)$ respectively. With this identification the vectors are

$$\lambda = \begin{bmatrix} 0 \\ 0 \\ 0 \\ 3,000 \\ 100,000 \\ 5,000 \\ -600,000 \\ -8,000 \\ -10,000 \\ -600,000 \\ -\infty \\ -50,000 \end{bmatrix} \quad \text{and} \quad \Lambda = \begin{bmatrix} 8,000 \\ \infty \\ 6,000 \\ 600,000 \\ 600,000 \\ 10,000 \\ -300,000 \\ 0 \\ 0 \\ 60,000 \\ 0 \\ 0 \end{bmatrix}.$$

Recall here that balances of asset accounts are normally positive and those of liability and equity accounts are negative. Here are some comments on the restrictions implied by these vectors.

1. The a_1-balance is bounded by $0 and $8,000 because it is not possible to have a negative cash balance for account a_1, and, according to the company policy, cash balances over $8,000 are not permitted.

2. Bank account a_2 is bounded by 0 and ∞ because the company policy does not permit the bank account to be overdrawn and there is no upper limit for the balance of funds in the account.

3. According to the company's security policy, the trade debtors account is restricted and cannot exceed $6,000.

4. The assets represented by accounts a_4 and a_5 require a minimum investment to keep the equipment in working order, namely $3,000 and $100,000, respectively. In practice one would expect that the investment in these assets could be greater than the minimum, thereby improving the production process, but there is an upper limit imposed by company policy of $600,000 for both assets.

5. The balance of the inventory a_6 is bounded by a minimum amount $5,000, which is necessary to satisfy the orders from clients, and a maximum amount $10,000 imposed by the physical restrictions of the warehouse.

6. As regards share capital a_7, according to the legal status of the company (public corporation, limited corporation, limited partnership, etc.), there is a minimum share capital, in this case, $300,000. Moreover, the company can allow increases in capital up to $600,000. (Allow for signs in comments 6 – 10).

7. Taking into account the fact that the trade creditors represent a credit when purchasing inventories, the company's image requires that the balance of account a_8 be bounded, say by $8,000.

8. Just like the trade creditors account, it is inappropriate for the company to exceed a limit in the loans received account a_9, say the limit is $10,000. There could be various reasons for this restriction: lack of confidence at the bank, deterioration of certain ratios, etc.

9. As in the case of a_4, there is an upper bound of $600,000 for the balance of the depreciation account a_{10}. In addition it is usual for the company to depreciate a percentage of the a_4-balance per year, say 10%, so there is a lower limit of $60,000.

10. Finally, it is possible that, depending on the profit or loss situation, the company might wish to increase retained earnings, which explains the bound of $50,000 in a_{12} and unlimited balance in a_{11}.

The set B of allowable balance vectors for the system as defined in terms of the functions λ and Λ is therefore

$$B = \{\mathbf{b} \in \mathrm{Bal}_{12}(\mathbb{Z}) \mid \lambda(a_i) \leq b_i \leq \Lambda(a_i), \ i = 1, 2, \ldots, 12\}.$$

The extended model also has the capacity to generate reports. In this case, because of the small size of the company, it is assumed that there is a single report which corresponds to the equivalence relation E on the account set A with the following equivalence classes:

- $\{a_1, a_2, a_3, a_6\}$, which will be called "current assets".

- $\{a_4, a_5, a_{10}\}$, which will be called "non-current (tangible) assets".

- $\{a_7, a_{11}, a_{12}\}$, which will be labeled "capital and reserves".

- $\{a_8\}$: "current liabilities".

- $\{a_9\}$: "non-current liabilities".

Thus the set of reports is simply

$$\mathcal{E} = \{E\}.$$

According to the rules and policies of the company the authorizations needed for transactions affecting the various accounts are encoded in the following 12×5 control matrices – recall that C^+ displays authorizations needed for debits and C^- those for credits:

$$C^+ = \begin{bmatrix} 0 & 0 & 0 & 0 & 0 \\ 0 & 0 & 0 & 0 & 0 \\ 0 & 0 & 1 & 0 & 2 \\ 2 & 0 & 0 & 0 & 1 \\ 2 & 0 & 0 & 0 & 1 \\ 3 & 2 & 0 & 1 & 0 \\ 2 & 0 & 0 & 0 & 1 \\ 2 & 2 & 0 & 0 & 1 \\ 2 & 0 & 0 & 0 & 1 \\ 0 & 0 & 0 & 0 & 1 \\ 0 & 0 & 0 & 0 & 1 \\ 0 & 0 & 0 & 0 & 0 \end{bmatrix}, \quad C^- = \begin{bmatrix} 1 & 0 & 0 & 0 & 2 \\ 1 & 0 & 0 & 0 & 2 \\ 0 & 0 & 0 & 0 & 0 \\ 0 & 0 & 0 & 0 & 1 \\ 0 & 0 & 0 & 0 & 1 \\ 3 & 0 & 2 & 1 & 0 \\ 0 & 0 & 0 & 0 & 1 \\ 0 & 2 & 0 & 0 & 1 \\ 0 & 0 & 0 & 0 & 1 \\ 0 & 0 & 0 & 0 & 1 \\ 0 & 0 & 0 & 0 & 1 \\ 0 & 0 & 0 & 0 & 0 \end{bmatrix}.$$

In order to understand the role played by the matrices C^+ and C^-, we recall the significance of their entries. The columns of the matrices represent the units of the company α, β_1, β_2, γ and δ, while the rows represent the accounts $a_1, a_2, \ldots, a_{12}$.

- A debit in accounts a_1 (cash) or a_2 (bank) represents an injection of cash into the company and requires no authorization and so rows 1 and 2 of matrix C^+ consists of zeros. However, a transaction that credits a_1 or a_2 has to be approved by α and δ in that precise order.

- A debit in account a_3 (trade debtors), as indicated by C^+, requires two authorizations: first from department β_2 and then from department δ (in this order). A credit to this account requires no authorizations.

- A debit to accounts a_4 (machinery, plant and tools), a_5 (buildings and other structures), a_7 (share capital) or a_9 (loans received) requires that departments δ (accounts) and α (cash) agree in that order. On the other hand, a debit to account a_8 requires authorization first by δ (accounts department), α (cash) and β_1 (purchasing department), because some of the purchased inventories might be defective: however, the authorizations from α and β_1 can be given in any order. On the other hand, a credit to any of these accounts only requires the agreement of δ, except for a_8, which must also be approved by β_1.

- Account a_6 needs three authorizations for debits, in order from γ (warehouse), β_1 (purchasing department) and α (cash): the case of a credit is analogous, but β_1 must be changed to β_2 (sales department).

- Accounts a_{10} and a_{11} need only authorizations from δ (accounts department) for both debits and credits.

- Finally, account a_{12} requires no authorizations because its balance is the consequence of those of other accounts.

Next the frequency function, $\varphi : T_1 \to \mathbb{N} \cup \{\infty\}$ must be specified; it is identified with a 3-column vector with entries in $\mathbb{N} \cup \{\infty\}$, where the components of the vector are the values of φ. We have seen that for the company the set T_1 has just three elements, namely

- $\mathbf{w}_1 = $ depreciation over one year;

- $\mathbf{w}_2 = $ quarterly payments on a loan;

- $\mathbf{w}_3 = $ monthly payments of salaries and wages.

Thus the values of φ written in column form are

$$\varphi = \begin{bmatrix} 1 \\ 4 \\ 14 \end{bmatrix},$$

where $\varphi(\mathbf{w}_1) = 1$ since $\mathbf{w}_1$ is the depreciation over one whole year, and $\varphi(\mathbf{w}_2) = 4$ because the loan amortization is assumed to be quarterly; finally, $\varphi(\mathbf{w}_3) = 14$ because each worker has to receive his/her salary monthly with two extra payments at the middle and end of the year.

10.2. The Operation of the Model

We will now show how the 10-tuple model records some typical accounting activities of the company. Suppose that the initial (allowable) balance vector $\mathbf{b}_0$ and frequency counter κ_0 are given by

$$\mathbf{b}_0 = \begin{bmatrix} 7,000 \\ 14,000 \\ 3,000 \\ 100,000 \\ 200,000 \\ 7,000 \\ -300,000 \\ -2,000 \\ -5,000 \\ -5,000 \\ -10,000 \\ -9,000 \end{bmatrix}, \quad \kappa_0 = \begin{bmatrix} 0 \\ 2 \\ 7 \end{bmatrix}.$$

The components of $\mathbf{b}_0$ are the initial balances of the various accounts of the company, while κ_0 shows the frequencies with which the specific allowable transactions $\mathbf{w}_1, \mathbf{w}_2, \mathbf{w}_3$ have already been applied. Thus the initial state of the automaton $\mathcal{M}_A$ is $(\mathbf{b}_0, \kappa_0)$.

The transactions

Let us examine the effect of applying a chain of six allowable transactions $\mathbf{v}_1, \mathbf{v}_2, \ldots, \mathbf{v}_6$. The first three are of allowable types in T_0 and the remaining three are specific transactions from T_1. Keep in mind that only transactions in T_1 change the frequency counter. Each transaction is accompanied by two control matrices detailing the authorizations received from the various units. Thus a typical input is $(\mathbf{v}_i, M_i^+, M_i^-)$. Recall that we prefer to work with the reduced forms of these matrices M_i^{+*}, M_i^{-*}, in which zero rows corresponding to entries of $\mathbf{v}_i$ which are non-positive or non-negative respectively are omitted.

1. Transaction $\mathbf{v}_1$: purchase of \$1,000 inventories, \$150 to be paid

through bank and \$850 through trade creditors, is given by

$$\mathbf{v}_1 = \begin{bmatrix} 0 \\ -150 \\ 0 \\ 0 \\ 0 \\ 1,000 \\ 0 \\ -850 \\ 0 \\ 0 \\ 0 \\ 0 \end{bmatrix}.$$

In addition the transaction comes tagged by two control matrices which record the authorizations already obtained: in reduced form these are

$$M_1^{+*} = \begin{bmatrix} 3 & 2 & 0 & 1 & 0 \end{bmatrix}, \quad M_1^{-*} = \begin{bmatrix} 1 & 0 & 0 & 0 & 2 \\ 0 & 2 & 0 & 0 & 1 \end{bmatrix}.$$

Thus the row in M_1^{+*} is row 6 of M_1^{+} and the rows of M_1^{-*} are rows 2 and 8 of M_1^{-}. Notice that $M_1^{+*} >_{\mathbf{v}_1} C^{+}$ and $M_1^{-*} >_{\mathbf{v}_1} C^{-}$, which shows that M_1^{+} and M_1^{-} incorporate all the authorizations needed for the transaction $\mathbf{v}_1$. In this case $\mathbf{v}_1 \notin T_1$, so the frequency function does not change. Thus the new state of the system effected by the input $(\mathbf{v}_1, M_1^{+}, M_1^{-})$ is $(\mathbf{b}_1, \kappa_1)$ where $\mathbf{b}_1 = \mathbf{b}_0 + \mathbf{v}_1$ and $\kappa_1 = \kappa$. Hence

$$\mathbf{b}_1 = \begin{bmatrix} 7,000 \\ 13,850 \\ 3,000 \\ 100,000 \\ 200,000 \\ 8,000 \\ -300,000 \\ -2,850 \\ -5,000 \\ -5,000 \\ -10,000 \\ -9,000 \end{bmatrix} \quad \text{and} \quad \kappa_1 = \begin{bmatrix} 0 \\ 2 \\ 7 \end{bmatrix}.$$

The corresponding output is the report

$$[\mathbf{b}_1]_E = \begin{bmatrix} 31,850 \\ 295,000 \\ -319,000 \\ -2,850 \\ -5,000 \end{bmatrix}.$$

2. Transaction $\mathbf{v}_2$: sale of \$3,000 of inventories, with \$1,000 of the proceeds to be paid into cash and \$2,000 to trade debtors.

$$\mathbf{v}_2 = \begin{bmatrix} 1,000 \\ 0 \\ 2,000 \\ 0 \\ 0 \\ -3,000 \\ 0 \\ 0 \\ 0 \\ 0 \\ 0 \\ 0 \end{bmatrix}.$$

The transaction is tagged by control matrices

$$M_2^{+*} = \begin{bmatrix} 0 & 0 & 0 & 0 & 0 \\ 0 & 0 & 1 & 0 & 2 \end{bmatrix}, \quad M_2^{-*} = \begin{bmatrix} 3 & 0 & 2 & 1 & 0 \end{bmatrix}.$$

Since $M_2^{+*} >_{\mathbf{v}_2} C^+$ and $M_2^{+*} >_{\mathbf{v}_2} C^+$, the authorization process is complete. Also $\mathbf{v}_2 \notin T_1$, so the frequency function does not change. The new state is $(\mathbf{b}_2, \kappa_2)$ where

$$\mathbf{b}_2 = \begin{bmatrix} 8,000 \\ 13,850 \\ 5,000 \\ 100,000 \\ 200,000 \\ 5,000 \\ -300,000 \\ -2,850 \\ -5,000 \\ -5,000 \\ -10,000 \\ -9,000 \end{bmatrix} \quad \text{and} \quad \kappa_2 = \kappa_1 = \begin{bmatrix} 0 \\ 2 \\ 7 \end{bmatrix}.$$

The output is the report

$$[\mathbf{b}_2]_E = \begin{bmatrix} 31,850 \\ 295,000 \\ -319,000 \\ -2,850 \\ -5,000 \end{bmatrix} :$$

note that the output report has not changed, i.e., $[\mathbf{b}_1]_E = [\mathbf{b}_2]_E$, because transaction $\mathbf{v}_2$ has moved funds among accounts belonging to the same equivalence class, viz "current assets".

3. Transaction $\mathbf{v}_3$: a bank gives a loan of \$1,000 to purchase machinery.

$$\mathbf{v}_3 = \begin{bmatrix} 0 \\ 0 \\ 0 \\ 1,000 \\ 0 \\ 0 \\ 0 \\ 0 \\ -1,000 \\ 0 \\ 0 \\ 0 \end{bmatrix}.$$

In this case the transaction is tagged by the control matrices

$$M_3^{+*} = \begin{bmatrix} 2 & 0 & 0 & 1 & 0 \end{bmatrix}, \quad M_3^{-*} = \begin{bmatrix} 0 & 0 & 0 & 0 & 1 \end{bmatrix},$$

which list the authorizations received. However, in this case M_3^{+*} does not contain all the authorizations needed to approve the transaction v_3, because the accounts department has not authorized the purchase of machinery, despite the fact that it has authorized receipt of the bank loan. Therefore the transaction is rejected and the state remains $(b_3, \kappa_3) = (b_2, \kappa_2)$. Thus

$$b_3 = \begin{bmatrix} 8,000 \\ 13,850 \\ 5,000 \\ 100,000 \\ 200,000 \\ 5,000 \\ -300,000 \\ -2,850 \\ -5,000 \\ -5,000 \\ -10,000 \\ -9,000 \end{bmatrix},$$

and the output will be an error message.

4. Transaction v_4: depreciation of machinery, plant and tools by $5,000.

$$v_4 = \begin{bmatrix} 0 \\ 0 \\ 0 \\ 0 \\ 0 \\ 0 \\ 0 \\ 0 \\ 0 \\ -5,000 \\ 0 \\ +5,000 \end{bmatrix}.$$

The authorization matrices are

$$M_4^{+*} = \begin{bmatrix} 0 & 0 & 0 & 0 & 0 \end{bmatrix}, \quad M_4^{-*} = \begin{bmatrix} 0 & 0 & 0 & 0 & 1 \end{bmatrix},$$

which are satisfactory. In this case $\mathbf{v}_4 = \mathbf{w}_1 \in T_1$, so the frequency counter changes. The new state of the system is $(\mathbf{b}_4, \kappa_4)$, where

$$\mathbf{b}_4 = \begin{bmatrix} 8,000 \\ 13,850 \\ 5,000 \\ 100,000 \\ 200,000 \\ 5,000 \\ -300,000 \\ -2,850 \\ -5,000 \\ -10,000 \\ -10,000 \\ -4,000 \end{bmatrix} \quad \text{and} \quad \kappa_4 = \begin{bmatrix} 1 \\ 2 \\ 7 \end{bmatrix}.$$

The output is the report

$$[\mathbf{b}_4]_E = \begin{bmatrix} 31,850 \\ 290,000 \\ -314,000 \\ -2,850 \\ -5,000 \end{bmatrix}.$$

5. Transaction $\mathbf{v}_5$: quarterly loan amortization, with interest of $60 paid and $150 of principal repaid.

$$\mathbf{v}_5 = \begin{bmatrix} 0 \\ -210 \\ 0 \\ 0 \\ 0 \\ 0 \\ 0 \\ 0 \\ 150 \\ 0 \\ 0 \\ 60 \end{bmatrix}.$$

The authorizations obtained are:

$$M_5^{+*} = \begin{bmatrix} 2 & 0 & 0 & 0 & 1 \\ 0 & 0 & 0 & 0 & 0 \end{bmatrix}, \quad M_5^{-*} = \begin{bmatrix} 1 & 0 & 0 & 0 & 2 \end{bmatrix},$$

which are in order. Here $\mathbf{v}_5 = \mathbf{w}_2 \in T_1$, so once again the frequency counter changes. The new state of the system is $(\mathbf{b}_5, \kappa_5)$ where

$$\mathbf{b}_5 = \begin{bmatrix} 8,000 \\ 13,640 \\ 5,000 \\ 100,000 \\ 200,000 \\ 5,000 \\ -300,000 \\ -2,850 \\ -4,850 \\ -10,000 \\ -10,000 \\ -3,940 \end{bmatrix} \quad \text{and} \quad \kappa_5 = \begin{bmatrix} 1 \\ 3 \\ 7 \end{bmatrix}.$$

The output is the report

$$[\mathbf{b}_5]_E = \begin{bmatrix} 31,640 \\ 290,000 \\ -313,940 \\ -2,850 \\ -4,850 \end{bmatrix}.$$

6. Transaction $\mathbf{v}_6$: the company pays the monthly salaries of its

two employees through the bank account.

$$\mathbf{v}_6 = \begin{bmatrix} 0 \\ -3,000 \\ 0 \\ 0 \\ 0 \\ 0 \\ 0 \\ 0 \\ 0 \\ 0 \\ 0 \\ 3,000 \end{bmatrix}.$$

The authorizations are

$$M_6^{+*} = \begin{bmatrix} 0 & 0 & 0 & 0 & 0 \end{bmatrix}, \quad M_6^{-*} = \begin{bmatrix} 1 & 0 & 0 & 0 & 2 \end{bmatrix};$$

these are in order. Here $\mathbf{v}_6 = \mathbf{w}_3 \in T_1$, so once again the frequency counter will change. The new state of the system is $(\mathbf{b}_6, \kappa_6)$ where

$$\mathbf{b}_6 = \begin{bmatrix} 8,000 \\ 10,640 \\ 5,000 \\ 100,000 \\ 200,000 \\ 5,000 \\ -300,000 \\ -2,850 \\ -4,850 \\ -10,000 \\ -10,000 \\ -940 \end{bmatrix} \quad \text{and} \quad \kappa_6 = \begin{bmatrix} 1 \\ 3 \\ 8 \end{bmatrix}.$$

The output is the report

$$[\mathbf{b}_6]_E = \begin{bmatrix} 28,640 \\ 290,000 \\ -310,940 \\ -2,850 \\ -4,850 \end{bmatrix}.$$

Thus the balances of the company's accounts after the six transactions have been applied are recorded in the vector b_6.

Application of an audit

To conclude the example, let us consider what happens when the audit facility described in Chapter 9 is applied. Assume the company has a legal status which requires a yearly audit of its accounts. The auditor decides to ask for the balances of the following company accounts:

- inventories a_6;

- bank account a_2;

- trade debtors a_3;

- trade creditors a_8.

After stratified sampling, the auditor chooses to investigate the following items:

- finished products A and B;

- accounts in banks X, Y and Z;

- trade debtors M and N;

- trade creditor Q.

First of all, according to the accounting information, the final balance for inventories of $5,000 includes $1,200 for finished product A and $900 for finished product B. The auditor requires a count and proceeds to examine the documentation for both products in the warehouse. As a result of the investigation, the values of the products in the warehouse are found to be $1,500 for A and $850 for B. The difference of $50 in the value of product B is not considered to be significant, but the auditor decides to order an adjustment for product A. This records a debit to inventory in respect of product A amounting to $300, which implies a profit increase of the same amount. The corrected transaction vector is

$$\mathbf{v}_1^c = \begin{bmatrix} 0 \\ 0 \\ 0 \\ 0 \\ 0 \\ 300 \\ 0 \\ 0 \\ 0 \\ 0 \\ 0 \\ -300 \end{bmatrix}.$$

Observe that in this case the original transaction $\mathbf{v}_1$ equals $\mathbf{0}$ since there was no previous transaction involving the surplus of \$300 corresponding to the finished good A. Thus the automaton simply applies the transaction vector $\mathbf{v}_1^c - \mathbf{v}_1 = \mathbf{v}_1^c$ to the system in order to correct the error.

Secondly, assume that banks X, Y and Z supply lists of all activities in their accounts for the company through December 31. When these activities are checked for X and Y, the auditor observes that all of them correspond to approved transactions involving other asset or liability accounts. However, the case of bank Z is different; the auditor finds a debit of \$800 corresponding to the sale of inventories which has been wrongly applied to the account of trade debtor N, instead of being paid into the bank account. In this case it is necessary to cancel the original transaction, which was represented by the balance vector

$$\mathbf{v}_2 = \begin{bmatrix} 0 \\ 0 \\ 800 \\ 0 \\ 0 \\ -800 \\ 0 \\ 0 \\ 0 \\ 0 \\ 0 \\ 0 \end{bmatrix},$$

and replace it by the correct transaction

$$\mathbf{v}_2^c = \begin{bmatrix} 0 \\ 800 \\ 0 \\ 0 \\ 0 \\ -800 \\ 0 \\ 0 \\ 0 \\ 0 \\ 0 \\ 0 \end{bmatrix}.$$

Observe that this pair of transactions can be replaced by the single transaction

$$\mathbf{v}_2^c - \mathbf{v}_2 = \begin{bmatrix} 0 \\ 800 \\ -800 \\ 0 \\ 0 \\ 0 \\ 0 \\ 0 \\ 0 \\ 0 \\ 0 \\ 0 \end{bmatrix},$$

which represents the error. The effect of the auditing automaton is to apply the vector $\mathbf{v}_2^c - \mathbf{v}_2$ to the system to correct the error, as described in Chapter 9.

Finally, all debits and credits in which the trade creditor Q is involved are found to correspond to amounts in bank account and inventories, so from this information the auditor concludes that no adjustments are required involving the trade creditor account. The audit is now concluded.

10.3. Concluding Remarks

Throughout this work our aim has been to show how the operation of the double entry accounting system can be elucidated by the introduction of concepts and methods from abstract algebra. The remarkably simple key idea is that of a balance vector, which is used to display the account balances of a company at any instant, and also to represent the transactions which modify balances when an economic event affects the company. Balance vectors have the advantage that the signs of the entries show whether the transaction in question debits or a credits an account. They also have natural mathematical interpretations, which open up the use of a range of standard techniques and constructions from algebra.

The principal achievement of the investigation has been the construction of an algebraic model which closely represents the workings of a real life accounting system. The result is the so-called 10-tuple model, which is capable of screening balances and incoming transactions for appropriateness, verifying authorizations for such transactions, scrutinizing frequency of application of transactions, generating reports and detecting errors.

The algebraic model is most convincing when it is viewed as an automaton in which the balances form part of the state. The inputs contain transactions that change the state, including the frequency counts, while outputs include reports that are generated for the benefit of shareholders, creditors, clients and the public.

In addition to balance vectors and automata, other algebraic objects that have played a useful role in describing the operations of accounting systems include graphs, digraphs and monoids: furthermore, integer programming algorithms are important in the detection and correction of errors. The standard algebraic notion of a quotient structure is exactly what is called for in the formulation of a report.

It should be emphasized that, despite these successes, our approach is necessarily limited in its scope. Inevitably the application of algebraic methods to accounting theory cannot extend beyond depiction of the purely mechanical aspects of the subject. Throughout this work accounting systems are regarded as deterministic systems whose actions are always predictable consequences of the rules governing the system.

On the other hand, economics, like most social sciences, deals with the behavior of vast numbers of individual members of complex populations, for whose study statistical methods may be more appropriate. For algebraic methods to be successful there must be clear rules and well defined objects of study. In addition, we do not attempt to address the philosophical aspects of accounting theory. Nevertheless, despite these disclaimers, it is the authors' belief that a convincing case has been made for the claim that abstract algebra has much to contribute to an understanding of the accounting process.

List of Mathematical Symbols

X, Y, Z : sets.

$\overline{X}$: set of words in an alphabet X.

$|X|$: number of elements in a finite set X.

$\mathrm{Fun}(X)$: set of all functions on a set X.

$X \backslash Y$: a difference set.

$\mathrm{Cr}_{\lambda \in \Lambda}\, A_\lambda$: a cartesian product.

$\mathbb{P}, \mathbb{N}, \mathbb{Z}, \mathbb{Q}, \mathbb{R}$: respective sets of positive integers, natural numbers, integers, rational numbers, real numbers.

$\mathbf{b}, \mathbf{u}, \mathbf{v}$: column vectors.

$\mathrm{sppt}(\mathbf{v})$: support of a vector.

R^n : set of all n-column vectors over an ordered domain R.

$\mathrm{Bal}_n(R)$: set of n-balance vectors over R.

$\mathrm{Trans}_n(R)$: set of n-transaction vectors over R.

$\mathbf{e}(i, j)$: balance vector with ith entry 1, jth entry -1 and other entries 0.

$\mathrm{type}(\mathbf{v})$: type of a balance vector $\mathbf{v}$.

$\tau_{\mathbf{v}} = \mathbf{v}'$: transaction which adds $\mathbf{v}$.

$f_{\mathbf{v}}$: function which adds $\mathbf{v}$ subject to allowability.

$\mathcal{A}$: an accounting system.

$\mathcal{A}_1 \vee \mathcal{A}_2 \vee \cdots \vee \mathcal{A}_k$: a join of accounting systems.

$\mathbf{v}^*$: an allowable vector in a join of accounting systems.

$\mathcal{A}/E$: a quotient accounting system.

$[x]_E$: E-equivalence class of x.

$\mathbf{v}_E$: a vector in the quotient system $\mathcal{A}/E$.

σ_E : canonical epimorphism associated with equivalence relation E.

θ^* : homomorphism induced by the function θ.

$\text{Mon}\langle X \rangle$: submonoid generated by X.

$\text{Mon}(\mathcal{A})$: monoid of an accounting system $\mathcal{A}$.

$\mathcal{S} = (Z, X, \delta)$: a semiautomaton.

$\mathcal{M} = (Z, X, Y, \delta, \lambda)$: an automaton.

$\mathcal{M}_{\mathcal{A}}$, $\mathcal{T}_{\mathcal{A}}$: automaton and time enhanced automaton of an accounting system $\mathcal{A}$.

$\text{Im}(\theta)$: image of a function/homomorphism.

$\text{Ker}(\theta)$: kernel of a homomorphism.

$M \oplus N$: a direct sum of modules.

A^T : transpose of a matrix.

C^+, C^- : control matrices.

M^{+*}, M^{-*} : reduced control matrices.

$E(i, j)$: matrix with 1 as the (i, j) entry and other entries 0.

$M_n(R)$: set of $n \times n$ matrices over R.

$V_{is}(D), V_{so}(D), V_{si}(D), V_c(D)$: sets of vertices of a digraph D.

$\binom{n}{r}$: a binomial coefficient.

$[\ell]$: greatest integer less than or equal to ℓ.

$S(k, m)$: a Stirling number of the second kind.

Bibliography

Mathematics References

[1] Biggs, N.L. Discrete Mathematics, 2nd ed. Oxford. 2002.

[2] Brualdi, R.A. Introductory Combinatorics, 5th ed. Prentice-Hall, Upper Saddle River, NJ. 2010.

[3] Cooper, S.B. Computability Theory. Chapman Hall, Boca Raton, FL. 2004.

[4] Hennie, F.C. Introduction to Computability. Addison-Wesley, Reading, MA. 1977.

[5] Hopcroft, J. and Ullman, J. Introduction to Automata Theory, Languages and Computation. Addison-Wesley, Reading, MA. 1979.

[6] Kolman, B. and Beck, R.E. Elementary Linear Programming with Applications, 2nd ed. Academic Press, San Diego, CA. 1995.

[7] Lidl, R. and Pilz, G. Applied Abstract Algebra. Springer, New York. 1998.

[8] Robinson, D.J.S. An Introduction to Abstract Algebra. W. de Gruyter, Berlin. 2003.

[9] Robinson, D.J.S. A Course in Linear Algebra with Applications, 2nd ed. World Scientific, Singapore. 2006.

[10] Rosen, K.H. Discrete Mathematics and its Applications, 6th ed. McGraw-Hill, Boston, MA. 2007.

[11] Strang, G. Linear Algebra and its Applications, 3rd ed. Harcourt Brace Jovanovich, San Diego, CA. 1988.

[12] West, D.B. Introduction to Graph Theory, 2nd ed. Prentice-Hall, Upper Saddle River, NJ. 2001.

Accounting References

Ames, E. (1983). Automaton and group structures in certain economic adjustment mechanisms. Mathematical Social Sciences 6(2), 247-260.

Arya, A., Fellingham, J.C., Mittendorf, B. and Schroeder, D.A. (2004). Reconciling Financial Information at Varied Levels of Aggregation. Contemporary Accounting Research, 21(2), 303.

Arya, A., Fellingham, J.C. and Schroeder, D.A. (2000a). Accounting information, aggregation, and discriminant analysis. Management Science, 46(6), 790.

Arya, A., Fellingham, J.C. and Schroeder, D.A. (2000b). Estimating transactions given balance sheets and an income statement. Issues in Accounting Education, 15(3), 393.

Aukrust, O. (1955). Nationalregnskap-Teoretske Prinsipper (National Income Accounting Theoretical Principles), Oslo, Statistik Centralburå.

Aukrust, O. (1966). An Axiomatic Approach to National Accounting: An Outline. Review of Income and Wealth, 12(3), 179-190.

Balzer, W. and Mattessich, R. (1991). An axiomatic basis of accounting: a structuralist approach. Theory and Decision, 30, 213-243.

Balzer, W. and Mattessich, R. (2000). Formalizing the basis of accounting, in Balzer,W., Sneed, J.D. and Moulines, C.U. eds. Structuralist Knowledge Representation-Paradigmatic Examples, Amsterdam, Rodopi, Atlanta GA (Vol. 75 of the Poznan Studies in the Philosophy of the Sciences and Humanities), 99-126.

Barley, S.R. (1983). Semiotics and the Study of Occupational and Organizational Cultures. Administrative Science Quarterly, 28(3), 393-413.

Belkaoui, A. (1978). Linguistic Relativity in Accounting. Accounting Organizations and Society, 3(2), 97-104.

Belkaoui, A. (1980a). The Impact of Socio-Economic Accounting Statements on the Investment Decision: An Empirical Study. Accounting, Organizations and Society, 5(3), 263-283.

Belkaoui, A. (1980b). The Interprofessional Linguistic Communication of Accounting Concepts: An Experiment in Sociolinguistics. Journal of Accounting Research, 18(2), 362-374.

Blackwell, D. (1951). Comparison of Experiments. In Proceedings of the Second Berkeley Symposium in Mathematical Statistics and Probability, edited by J. Neyman. Berkeley: University of California Press, 93-102.

Blackwell, D. (1953). Equivalent Comparison of Experiments. Annals of Mathematical Statistics, 24(2), 267-272.

Botafogo, F. (2009). Algebraic accounting: an introduction to accountancy's axiomatics. Working paper, São Paolo, Brazil.

Brewer, C. (1987). On the nature of accounting information sets. Typescript.

Butterworth, J.E. (1967). Accounting Systems and Management Decision: an Analysis of the Role of Information in the Managerial Decision Process. Unpublished Ph.D. Dissertation, University of California-Berkeley.

Cayley, A. (1894). The Principle of Bookkeeping by Double Entry. Cambridge University Press.

Chambers, R.J. (1966). Accounting, Evaluation and Economic Behaviour. Prentice-Hall. Englewood Cliffs, NJ. (Reprinted in Accounting Classics Series. Scholars Books Co., Houston, TX. 1975).

Cooke, T. and Tippett, M. (2000). Double entry bookkeeping, structural dynamics and the value of the firm. British Accounting Review, 32(3), 261-288.

Cruz Rambaud, S. and García Peréz, J. (2005). The accounting system as an algebraic automaton. International Journal of Intelligent Systems, 20, 827-842.

De Morgan, A. (1846). Elements of Arithmetic, 5th ed. Appendix, On the Main Principle of Book-Keeping. Taylor and Walton, London.

Demski, J.S. (1980). Information Analysis, 2nd ed. Addison-Wesley, Reading, MA.

Demski, J.S. (2007). Is accounting an academia discipline? Accounting Horizons, 21(2), 153-157.

Demski, J.S., Fitzgerald, S.A., Ijiri, Y. and Lin, H. (2006) Quantum information and accounting information: their salient features and conceptual applications. Journal of Accounting and Public Policy, 25, 435-464.

Demski, J.S., Fitzgerald, S.A., Ijiri, Y. and Lin, H. (2009). Quantum information and accounting information: exploring conceptual applications of topology. Journal of Accounting and Public Policy, 28, 133-147.

Demski, J.S., Patell, J.M. and Wolfson, M.A. (1984). Decentralized choice of monitoring systems, The Accounting Review, 59(1), 16-34.

Edwards, E.O. and Philip W.B. (1961). The Theory and Measurement of Business Income. University of California Press Berkeley, CA.

Ellerman, D. (1982). Economics, Accounting, and Property Theory. D.C. Heath, Lexington, MA.

Ellerman, D. (1985). The mathematics of double entry bookkeeping. Mathematics Magazine, 58, 226-233.

Ellerman, D. (1986). Double entry multidimensional accounting. Omega, International Journal of Management Science, 14(1), 13-22.

Fisher, I.E. (2004). On the structure of financial accounting standards to support digital representation, storage, and retrieval. Journal of Emerging Technologies in Accounting, 1(1), 23-40.

Fisher, I.E. and Garnsey, M.R. (2006). The semantics of change as revealed through an examination of financial accounting standards amendments. Journal of Emerging Technologies in Accounting, 3(1), 41-60.

Garnsey, M.R. and Fisher, I.E. (2008). Appearance of new terms in accounting language: a preliminary examination of accounting pronouncements and financial statements. Journal of Emerging Technologies in Accounting, 5(1), 17-36.

Gibbons, M. and Willett, R.J. (1997). A new light on accrual, aggregation and allocation, using an axiomatic analysis of accounting. Abacus, 33(2), 137-168.

Gjesdal, F. (1981). Accounting for stewardship. Journal of Accounting Research, 19(1), 208-231.

Hamilton, W.R. (1837). Theory of conjugate functions, or algebraic couples: with a preliminary and elementary essay on algebra as the science of pure time. Transactions of the Royal Irish Academy 17, 293-422.

Husserl, E. (1931). Ideas. Allen and Unwin, London.

Ijiri, Y. (1967). The Foundations of Accounting Measurement: a Mathematical, Economic and Behavioral Inquiry. Prentice-Hall, Englewood Cliffs, NJ.

Ijiri, Y. (1975). Theory of Accounting Measurement. American Accounting Association, Sarasota, FL.

Lebar, M.A. (1982) A general semantics analysis of selected sections of the 10-K, the annual report to shareholders, and the financial press release. The Accounting Review, 57(1), 176-189.

Mattessich, R. (1957). Towards a general and axiomatic foundation of accountancy. Accounting Research, 8, 328-355.

Mattessich, R. (1964). Accounting and Analytical Methods. Irwin, Homewood.

Mattessich, R. (1995). Critique of Accounting–Examination of the Foundations and Normative Structure of Accounting. Quorum-Books, Greenwood Publishing Group, Westport, CT.

Mattessich, R. (1998). From accounting to negative numbers: A signal contribution of Medieval India to mathematics. Accounting Historians Journal, 25(2), 129-145.

Mattessich, R. (2000). The Beginnings of Accounting and Accounting Thought–Accounting Practice in the Middle East (8000 B.C. to 2000 B.C.) and accounting thought in India (300 B.C. and the Middle Ages). Garland Publishing, New York, NY.

Mattessich, R. (2003). Accounting research and researchers of the nineteenth century and the beginning of the twentieth century: an international survey of authors, ideas and publications. Accounting, Business and Financial History, 13(2), 171-205.

Mattessich, R. (2005a). The information economic perspective of accounting – its coming of age. Accounting Working Paper, Sauder School of Business, University of British Columbia.

Mattessich, R. (2005b). A Concise History of Analytical Accounting: Examining the use of Mathematical Notions in our Discipline. Spanish Journal of Accounting History, 2, 123-153.

Mattessich, R. and Galassi, G. (2000). History of the spreadsheet: from matrix accounting to budget simulation and computerization, in AECA ed., Accounting and History. Selected Papers from the 8th Congress of Accounting Historians, Madrid. Asociación Española de Contabilidad y Administración, 203-232.

McCloskey, D. (1983). The rhetoric of economics. Journal of Economic Literature, 21, 481-517.

McClure, M. (1983). Accounting as Language: a Linguistic Approach to Accounting. Unpublished Ph.D. Dissertation, University of Illinois.

Nehmer, R.A. (1988). Accounting Information Systems as Algebras and First Order Axiomatic Models. Unpublished Ph.D. Dissertation, University of Illinois.

Nehmer, R.A. (2010). Accounting systems as first order axiomatic models: consequences for information theory. International Journal of Mathematics in Operational Research, 2(1), 99-112.

Nehmer, R.A. and Robinson, D.J.S. (1997). An algebraic model for the representation of accounting systems. Annals of Operations Research, 71(1), 179-198.

Pacioli, L. (1963). Summa de Arithmetica, Geometria, Proportioni et Proportionalita: Distintio Nona, Tractus XI, Particularis de Computis et Scripturis. (1494). Translated by Brown, R.G., Johnston, K.S., as "Pacioli on Accounting", McGraw-Hill, New York, NY.

Paton, W.A. (1922). Accounting Theory. Ronald Press, New York, NY. (Reprinted by Accounting Studies Press, Chicago, IL. 1962).

Stephens, R.G., Dillard, J.F. and Dennis, D.K. (1985). Implications of formal grammars for accounting policy development. Journal of Accounting and Public Policy, 4, 123-148.

Tippett, M. (1978). Axioms of accounting measurement. Accounting and Business Research, Autumn, 266-278.

Tyrvainen, P., Kipelainen, T. and Jarvenpaa, M. (2005). Patterns and measures of digitalisation in business unit communication. International Journal of Business Information Systems, 1, 199-219.

Velupillai, K.V. (2005). The unreasonable ineffectiveness of mathematics in economics. Cambridge Journal of Economics, 29, 489-872.

Willett, R.J. (1987). An axiomatic theory of accounting measurement. Accounting and Business Research, 17, 155-171.

Willett, R.J. (1988). An axiomatic theory of accounting measurement - Part II. Accounting and Business Research, 19, 79-91.

Willett, R.J. (1991). Theory of accounting measurement structures. IMA Journal of Management Mathematics, 3(1), 45-59.

Index